Beiträge zur Graphischen Datenverarbeitung

Editor:
Zentrum für Graphische Datenverarbeitung e.V. Darmstadt (ZGDV)

J. L. Encarnação H.-O. Peitgen
G. Sakas G. Englert (Eds.)

Fractal Geometry and Computer Graphics

With 172 Figures

Springer-Verlag
Berlin Heidelberg New York
London Paris Tokyo
Hong Kong Barcelona
Budapest

Series Editor

ZGDV, Zentrum für Graphische Datenverarbeitung e. V.
Wilhelminenstraße 7, W-6100 Darmstadt, FRG

Editors

José L. Encarnação
Georgios Sakas
Technische Hochschule Darmstadt
Institut für Informationsverwaltung und Interaktive Systeme
Wilhelminenstraße 7, W-6100 Darmstadt, FRG

Heinz-Otto Peitgen
Institut für Dynamische Systeme
Universität Bremen, W-2800 Bremen 33, FRG

Gabriele Englert
Fraunhofer Gesellschaft
Institut für Graphische Datenverarbeitung
Wilhelminenstraße 7, W-6100 Darmstadt, FRG

ISBN 3-540-55317-7 Springer-Verlag Berlin Heidelberg New York
ISBN 0-387-55317-7 Springer-Verlag New York Berlin Heidelberg

Library of Congress Cataloging-in-Publication Data
Fractal geometry and computer graphics / J. L. Encarnaçao ... [et al.], (eds.).
p. cm. – (Beiträge zur graphischen Datenverarbeitung) Includes bibliographical references and index.
ISBN 3-540-55317-7 (Berlin). – ISBN 0-387-55317-7 (New York)
1. Fractals. 2. Computer graphics. I. Encarnaçao, José Luis. II. Series.
QA614.86.F68 1992 514'.74–dc20 92-10008

Printed in Germany

Cover picture shows turbulent motion of 2-D clouds simulating condensation see page 192, Fig. 67 by G. Sakas
Typesetting: Camera ready by author
33/3140-5 4 3 2 1 0 - Printed on acid-free paper

Foreword

Fractal Geometry and Computer Graphics have been in the past two areas of research developing independent from each other. They have developed their mathematics, algorithms, techniques, and applications with their own priorities and addressing the issues relevant to their own, specific development. Now, based on the hardware developments (microelectronics, workstations, etc.) and driven by application needs, these two disciplines are getting more and more a common interface, where both integrate the know-how, the technologies and the methodologies coming from both sides. This book is a collection of papers addressing this issue. The nineteen papers published are structured around four main topics:

- Fundamentals,
- Computer Graphics,
- Simulation,
- Picture Analysis,

and are a first attempt to collect statements, views and first experiences with *basics, interrelationships and applications of* **Fractal Geometry and Computer Graphics**. We hope that this book is a first, very promising start in developing synergies and interrelationships between these two disciplines.

Included in part V are two reports on the results of discussions which took place in working groups during the workshop. These reports give **surveys and perspectives** on

- modeling and simulation, and
- random fractals.

Here we tried to collect opinions, open issues, and trends, which we hope will be the basis for intensive research in the near future.

I would like to thank H.-O. Peitgen for co-chairing the workshop, all participants for their contributions, and specially G. Englert and G. Sakas for all the effort they had in the very successful organization and implementation of the workshop, as well as in co-editing the book reporting on it.

Darmstadt, November 1991 J.L. Encarnação

Contents

III. Part: Simulation

IV. Part: Picture Analysis

V. Part: Working Group Results

List of Authors

Dr. H.R. Bittner
Pfarrer-Bleymann-Gasse 7
W-8730 Bad Kissingen - Poppenroth

Prof. Dr. W. Düchting
Universität Siegen
FB 12
Hölderlinstr. 3
W-5900 Siegen

Prof. Dr. J.L. Encarnação
GRIS (Graphisch-Interaktive Systeme)
FB 20 der TH Darmstadt
Wilhelminenstr. 7
W-6100 Darmstadt

Gabriele Englert
FhG-AGD
Fraunhofer-Arbeitsgruppe für
Graphische Datenverarbeitung
Wilhelminenstr. 7
W-6100 Darmstadt

P. Guillattaud
Centre de Recherche Paul Pascal
CNRS
Domaine Universitaire
F-33405 Talence Cedex

A. Haase
Universität Würzburg
Physikalisches Institut
Am Hubland
W-8700 Würzburg

F. v. Haeseler
Universität Bremen
Institut für Dynamische Systeme
Bibliothekstr. 1
W-2800 Bremen 33

M.S. Hammel
University of Calgary
MS 247
2500 University CA NW
Calgary AB T2N 1N4
Canada

Dr. P. Hanusse
Centre de Recherche Paul Pascal
CNRS
Domaine Universitaire
F-33405 Talence Cedex

Dr. H. Jürgens
Universität Bremen
Institut für Dynamische Systeme
Bibliothekstr. 1
W-2800 Bremen 33

Prof. G. Mantica
Service de Physique Theorique
Laboratoire de l' Institut de
Recherche Fondamentale du
commisariat a l' Energie Atomique
CEN-SACLAY
F-91191 Gif-sur-Yvette Cedex

U. Müssigmann
Fraunhofer-Institut IPA
Postfach 800 469
W-7000 Stuttgart 80

A. Odgaard
Orthopaedic Hospital Arhus
8000 Arhus
Dänemark

Prof. Dr. H.-O. Peitgen
Universität Bremen
Institut für Dynamische Systeme
Bibliothekstr. 1
W-2800 Bremen 33

Prof. Dr. P.J. Plath
Universität Bremen
FB Chemie
Postfach 33 04 40
W-2800 Bremen 33

Prof. P. Prusinkiewicz
University of Calgary
MS 247
2500 University CA NW
Calgary AB T2N 1N4
Canada

Dr. W. Rümelin
MSG Marine- und
Sondertechnik GmbH
MT 731
Hünefeldstr. 1-5
W-2800 Bremen 1

G. Sakas
GRIS (Graphisch-Interaktive Systeme)
FB 20 der TH Darmstadt
Wilhelminenstr. 7
W-6100 Darmstadt

M. Schendel
FhG-AGD
Fraunhofer-Arbeitsgruppe für
Graphische Datenverarbeitung
Wilhelminenstr. 7
W-6100 Darmstadt

J. Schwietering
Leher Heerstr. 17
W-2800 Bremen 33

Prof. Dr. M. Sernetz
Justus-Liebig-Universität
Inst. f. Biochemie u. Endokrinologie
Frankfurter Str. 100
W-6300 Gießen

G. Skordev
Universität Bremen
Institut für Dynamische Systeme
Bibliothekstr. 1
W-2800 Bremen 33

Prof. Dr. J.H. Spurk
TH Darmstadt
FB 16, FG Techn. Strömungslehre
Petersenstr. 30
W-6100 Darmstadt

Dr. Elisabeth Swart
Universität Bremen
FB 2
Postfach 33 04 40
W-2800 Bremen 33

Dr. Jutta Syha
Universität Würzburg
Physikalisches Institut
Am Hubland
W-8700 Würzburg

Prof. Dr. R.F. Voss
IBM Research
P.O. Box 218
Yorktown Heights, NY 10598
USA

P. Wlczek
Justus-Liebig-Universität
Inst. f. Biochemie u. Endokrinologie
Frankfurter Str. 100
W-6300 Gießen

Cornelia Zahlten
Universität Bremen
Institut für Dynamische Systeme
Bibliothekstr. 1
W-2800 Bremen 33

I. Part: Fundamentals

Linear Cellular Automata, Substitutions, Hierarchical Iterated Function Systems and Attractors

F. v. Haeseler, H.-O. Peitgen, G. Skordev

Institut für Dynamische Systeme, Universität Bremen

1 Introduction

Many constructions and algorithms from fractal geometry have become fundamental tools within the *naturalism program* of computer graphics. As the interest of computer graphics is shifting towards new frontiers such as scientific visualization some other basic concepts from fractal geometry come into focus. Cellular automata (CA) in particular are becoming a premier modelling and simulation tool in engineering and the basic sciences. CA live in a discrete world and are local in nature. They produce structure and patterns subject to a set of rules (look-up table) which determine the state of a growing cell from the state of its neighbors.

One of the major theoretical and practical problems is concerned with the *global CA problem*, i.e. with the pattern formation which evolves as a CA produces for a very long time. The nature of this important problem has been analyzed only for very special cases so far. For example, it is known that certain look-up tables generate Sierpinski gasket-like patterns [WIL84], [WOL83]. In general, however, it has been an open problem to understand the global pattern formation of a CA. The problem has several aspects. The first one which comes to mind is the problem of proper scaling, or, in what sense can one discuss an infinitely grown pattern? This problem can have several solutions. A very elegant one was proposed by S. Wilson in [WIL84]. The second problem is then to decipher the global pattern formation.

Our note solves these theoretical problems for the class of all (one-dimensional) linear cellular automata (LCA) by showing that the global patterns of a LCA can be understood as the attractor of an associated iterated function systems (IFS). The solution is presented in several steps. First we will embed LCA into a more general class, the class of matrix substitution systems (MSSs). MSSs are rooted in the work of Mandelbrot [MAN82] and are often used to generate basic fractals such as the Cantor set, the Sierpinski triangle, and the Peano curve. But they are also related to L-systems, see [DEK82] for a rigorous mathematical treatment, and see [PRU90] for an application of L-systems in botany. Our matrix substitution systems are a slight generalization of a concept presented in [DEK90]. In the next step we will demonstrate that MSSs are in turn deeply related to iterated function systems.

Iterated function systems were made popular and studied in connection with image compression problems by Barnsley, e.g. [BAR85, BAR88]. The concept of an IFS was introduced by Williams [WIL71] and Hutchinson [HUT81]. They provide an alternative method to generate fractal sets which is more in the frame of dynamical systems theory. Meanwhile there exist several extensions of iterated function systems, e.g. recurrent iterated function systems [BAR89, BER88, BER89] or mixing iterated function systems [BA88a], and hierarchical iterated function systems (HIFS) [JUE91]. The first explicit use of the idea of hierarchical iterated function systems seems to be in [BED86]. In fact Bedford shows that L-systems can be understood through HIFSs, see also [BAN]. We will introduce here a special version of hierarchical iterated function systems which we have used successfully to decipher the geometric patterns in Pascal's triangle in [HAE91]. These HIFSs are based on an n-adic description and we call them n-adic HIFSs. This description leads to elementary number theoretical characterizations of the associated attractors. For Pascal's triangle, see [HAE91], this characterization is in fact equivalent to a classical number theoretical result of E.E. Kummer [KUM52], p. 115f.

With regard to the global LCA problem we will discuss a mathematical algorithm which for a given LCA, or for that matter MSS, designs an n-adic HIFS whose attractor in some sense yields the properly rescaled global pattern of the given LCA, or MSS. Moreover, we will discuss several interesting consequences of this deep and amazing relation. One of these is a beautiful formula, found by S. Willson [WI87a], for the computation of the Hausdorff dimension of LCA which uses some earlier work from [MAU88]. Moreover, we shall also show that certain n-adic-HIFSs in turn induce a matrix substitution system. In other words, LCA are special MSSs, and MSSs and HIFs are closely related. Our result together with the beautiful result of Bedford [BED86], which shows that L-systems and HIFs are also related, suggests that HIFs are a natural frame for deterministic fractals.

2 Matrix Substitution Systems

Substitutions systems provide an elegant concept to generate fractal sets, cf. [DEK89, DEK90, MAN85]. Substitutions are transformations which assign words, i.e. concatenations of symbols, to words. Together with a graphical interpretation of words and a proper rescaling one obtains often fractal sets.

In this section we describe a simple concept of substitution. For the sake of simplicity we restrict ourselves to the case of generating fractals in $\mathbb{R}^2$.

Let $V = \{v_0, \ldots, v_N\}$ be a finite set, the set of symbols, together with a distinguished element v_0. The set of words is then given by

$$\mathcal{M} = \left\{ \omega : \mathbb{N}_0^2 \to V \;\middle|\; \begin{array}{c} \text{there is } K \in \mathbb{N} \text{ such that} \\ \omega(l,m) = v_0 \text{ if } l \text{ or } m \geq K \end{array} \right\}$$

where $\mathbb{N}_0 = \mathbb{N} \cup \{0\}$. The set $\mathcal{M}$ may be considered as the subset of all (infinite) matrices with entries in V, such that allmost all entries are trivial, i.e. equal to

v_0.

We now fix $n \in \mathbb{N}, n \geq 2$ and define the subset $\mathcal{M}_n$ in $\mathcal{M}$ by

$$\mathcal{M}_n = \{\omega : \mathbb{N}_0^2 \to V \,|\, \omega(l,m) = v_0 \text{ for all } l,m \geq n\},$$

i.e. $\mathcal{M}_n$ is the set of all $n \times n$-matrices in $\mathcal{M}$. By $\underline{v_0} \in \mathcal{M}$ we denote the mapping such that $\underline{v_0}(l,m) = v_0$ for all $(l,m) \in \mathbb{N}_0^2$. For $v \in V^* = V \setminus \{v_0\}$ we define $\underline{v}$ as

$$\underline{v}(l,m) \quad = \quad \begin{cases} v & \text{if } (l,m) = (0,0) \\ v_0 & \text{otherwise} \end{cases} \quad .$$

Definition 2.1 *A mapping $\sigma : V \to \mathcal{M}_n$ is called a* <u>*matrix substitution system*</u> *provided that*
1) $\sigma(v_0) = \underline{v_0}$, and
2) for any $v \in V^*$ there exists a sequence $(j_l, k_l)_{l \in \mathbb{N}} \subset \mathbb{N}_0^2$ such that

$$\sigma(\ldots\sigma(\sigma(\sigma(v)(j_1,k_1))(j_2,k_2))(j_3,k_3)\ldots)(j_l,k_l) \neq v_0$$

for all $l \in \mathbb{N}$.

Remark 2.1

Condition 2) of definition 2.1 implies that $\sigma(v) \not\equiv \underline{v_0}$ for all $v \in V^*$.

A matrix substitution system $\sigma : V \to \mathcal{M}_n$ induces a mapping $\Sigma : \mathcal{M} \to \mathcal{M}$, which is defined by

$$\Sigma(\omega) = \nu$$

where

$$\nu(l,m) = \sigma(\omega(r_1,r_2))(s_1,s_2)$$

and $l = r_1 n + s_1$ and $m = r_2 n + s_2$ with $0 \leq s_1, s_2 \leq n-1, r_1, r_2 \in \mathbb{N}_0$. In terms of matrices the mappings σ and Σ have the the following interpretation. Let A be a matrix with entries in V. Then replace every entry v_{ij} of A by the $n \times n$-matrix which is determined by $\sigma(v_{ij})$. This yields a new matrix B which is $\Sigma(A)$.

Remark 2.2

Let $\sigma : V \to \mathcal{M}_n$ be a matrix substitution system and $\Sigma : \mathcal{M} \to \mathcal{M}$ be the induced mapping. With Σ^k we denote the k-th iterate of Σ. Then 2) of definition 2.1 reads

$$\Sigma^k(\underline{v}) \not\equiv \underline{v_0}$$

for all $v \in V^*$ and all $k \in \mathbb{N}$.

The appropriate space for graphical representations of MSSs is $\mathcal{H}(\mathbb{R}^2)$ which denotes the metric space of all compact non-empty subsets of $\mathbb{R}^2$ equipped with the Hausdorff metric. Note that if the Hausdorff metric h is induced by the maximum metric (or any complete metric for that matter) on $\mathbb{R}^2$, then $\mathcal{H}(\mathbb{R}^2)$ is in turn a complete metric space.

For $(l,m) \in \mathbb{N}_0^2$ we denote by $I_{l,m}$ the set

$$I_{l,m} = \{(x,y) \in \mathbb{R}^2 \,|\, l \leq x \leq l+1, m \leq y \leq m+1\},$$

i.e. the closed unit square located at the lattice point $(l,m) \in \mathbb{N}_0^2$.

Definition 2.2 *The mapping $G : \mathcal{M} \to \mathcal{H}(\mathbb{R}^2) \cup \{\emptyset\}$ defined by*

$$G(\omega) = \bigcup_{\omega(l,m) \neq v_0} I_{l,m}$$

is called graphical representation of $\mathcal{M}$.

Remarks 2.3

1. $G(\omega) = \emptyset$ if and only if $\omega \equiv \underline{v_0}$.

2. If $g : V^* \to \mathcal{H}(\mathbb{R}^2)$ is any mapping, then g induces a graphical representation $\tilde{G}$ by

$$\tilde{G}(\omega) = \bigcup_{\omega(l,m) \neq v_0} g(\omega(l,m)) + (l,m),$$

where $g(\omega(l,m)) + (l,m)$ denotes the image of $g(\omega(l,m))$ under the translation $t(x,y) = (x+l, y+m)$ on $\mathbb{R}^2$.

Proposition 2.3 *Let $\sigma : V \to M_n$ be a matrix substitution system and Σ be the induced mapping on $\mathcal{M}$. Then for all $\omega \in \mathcal{M}$, $\omega \not\equiv \underline{v_0}$, the sequence*

$$\left(\frac{1}{n^k} G\Sigma^k(\omega)\right)_{k \in \mathbb{N}}$$

is a Cauchy sequence in $(\mathcal{H}(\mathbb{R}^2), h)$.

Proof : This follows directly from the fact that $\frac{1}{n} G\Sigma(\omega) \subset G(\omega)$, see figure 1. □

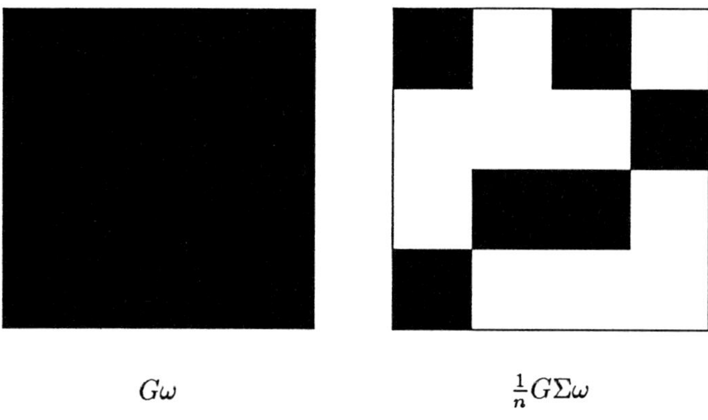

$G\omega$ $\qquad$ $\frac{1}{n} G\Sigma\omega$

Figure 1 : Visualization of the substitution

We denote the limit set of the sequence $\left(\frac{1}{n^k} G\Sigma^k(\omega)\right)$ by $\mathcal{A}(\omega)$.

Remarks 2.4

1. For $\omega \in \mathcal{M}\backslash\{\underline{v_0}\}$ there exists a natural decomposition of $\mathcal{A}(\omega)$ into subsets, where each subset is a translated limit set associated with a single symbol. This decomposition is given by

$$\mathcal{A}(\omega) = \bigcup_{\omega(l,m)\neq v_0} \mathcal{A}(\underline{\omega(l,m)}) + (l,m).$$

2. Let $g : V^* \to \mathcal{H}(\mathbb{R}^2)$ be any mapping and $\tilde{G} : \mathcal{M} \to \mathcal{H}(\mathbb{R}^2)$ the induced graphical representation. Then

$$\lim_{k\to\infty} \frac{1}{n^k}\tilde{G}\Sigma^k(\omega) = \mathcal{A}(\omega),$$

i.e. $\mathcal{A}(\omega)$ does not depend on the graphical representation.

Since the limit set $\mathcal{A}(\omega)$ is a composition of the limit sets $\mathcal{A}(\underline{v}), v \in V^*$, it is natural to study these basic building blocks more closely.

We will see that the n-adic expansion of real numbers provides a way to characterize the points of $\mathcal{A}(\underline{v})$ as points in $\mathbb{R}^2$ whose components have a particular n-adic expansion.

Proposition 2.4 *Let $\sigma : V \to \mathcal{M}_n$ be a matrix substitution system, and let $v \in V^*$. Then $\mathcal{A}(\underline{v})$ is the set of points $(x, y) \in I_{0,0}$ such that x and y have an n-adic expansion*

$$x = \textstyle\sum_{j=1}^{\infty} \frac{\alpha_j}{n^j}, \quad y = \sum_{j=1}^{\infty} \frac{\beta_j}{n^j}, \quad \alpha_j, \beta_j \in \{0, \ldots, n-1\} \tag{1}$$

which satisfies

$$\sigma(\ldots\sigma(\sigma(\underline{v})(\alpha_1, \beta_1))(\alpha_2, \beta_2)\ldots)(\alpha_l, \beta_l) \neq v_0 \tag{2}$$

for all $l \in \mathbb{N}$.

Proof : Let $(x, y) \in \mathcal{A}(\underline{v})$. We show that x and y satisfy (1) and (2). Since

$$\mathcal{A}(\underline{v}) = \bigcap_{k=0}^{\infty} \frac{1}{n^k} G\Sigma^k(\underline{v})$$

we have $n^k(x, y) \in G\Sigma^k(\underline{v})$. Therefore there exist natural numbers l_k, m_k such that $n^k(x, y) \in I_{l_k,m_k}, \frac{1}{n}I_{l_{k+1},m_{k+1}} \subset I_{l_k,m_k}$ and

$$\Sigma^k(\underline{v})(l_k, m_k) \neq v_0 \tag{3}$$

for all $k \in \mathbb{N}$. Now we choose $\alpha_1 = l_1$, $\beta_1 = m_1$ and $\alpha_k, \beta_k \in \{0, \ldots, n-1\}$ such that $l_k = nl_{k-1} + \alpha_k$, and $m_k = nm_{k-1} + \beta_k$. Then

$$x = \sum_{j=1}^{\infty} \frac{\alpha_j}{n^j} \text{ and } y = \sum_{j=1}^{\infty} \frac{\beta_j}{n^j}$$

is an n-adic expansion of x and y, and (3) becomes

$$\Sigma^k(\underline{v})(\alpha_1 n^{k-1} + \ldots \alpha_{k-1} n + \alpha_k, \beta_1 n^{k-1} + \ldots \beta_{k-1} n + \beta_k) =$$
$$\sigma(\Sigma^{k-1}(\underline{v})(\alpha_1 n^{k-2} + \ldots \alpha_{k-1}, \beta_1 n^{k-2} + \ldots \beta_{k-1}))(\alpha_k, \beta_k) \neq v_0$$

by the definition of Σ. Similar arguments show that a pair (x, y) which satisfies (1) and (2) yields coordinates of a point in $\mathcal{A}(\underline{v})$. □

We conclude this section with two matrix substitution systems each of which generates a classical fractal set.

Let $V = \{0, 1\}$, and define

$$\sigma(0) = \begin{pmatrix} 0 & 0 \\ 0 & 0 \end{pmatrix}, \sigma(1) = \begin{pmatrix} 1 & 1 \\ 1 & 0 \end{pmatrix},$$

i.e. σ defines a $\mathcal{M}_2$-substitution. Then $\mathcal{A}(\underline{1})$ is a Sierpinski triangle.

Let $V = \{0, 1\}$, and define a $\mathcal{M}_3$-substitution by

$$\sigma(0) = \begin{pmatrix} 0 & 0 & 0 \\ 0 & 0 & 0 \\ 0 & 0 & 0 \end{pmatrix}, \sigma(1) = \begin{pmatrix} 1 & 1 & 1 \\ 1 & 0 & 1 \\ 1 & 1 & 1 \end{pmatrix},$$

Then $\mathcal{A}(\underline{1})$ is the Sierpinski carpet.

3 Hierarchical Iterated Function Systems

So far we have seen how matrix substitution systems may lead to fractal limit sets if one uses a proper graphical representation and scaling. In this section we will establish a crucial relation to hierarchical iterated function systems for the limit sets $\mathcal{A}(\omega)$ of a matrix substitution system. We shall show that the limit sets $\mathcal{A}(\underline{v})$, $v \in V^*$, are given by some distinguished components of the attractor of an associated hierarchical iterated function system (HIFS).

We start with an introduction to HIFS which is adapted to our purposes. For a more general introduction we refer to [MAU88, EDG90].

Let $n \in \mathbb{N}, n \geq 2$ and let V be a finite set with at least two elements and distinguished element v_0. As before V^* denotes the set $V \setminus \{v_0\}$, and $I = I_{0,0}$ denotes the closed unit square in $\mathbb{R}^2$. Furthermore, we consider the contractions

$$f_{\alpha,\beta} : I \to I$$
$$(x, y) \mapsto \left(\frac{x+\alpha}{n}, \frac{y+\beta}{n}\right)$$

for all $(\alpha, \beta) \in \{0, \ldots, n-1\}^2$, and to each pair $(v, w) \in V^* \times V^*$ we associate a subset $J(v, w) \subset \{0, \ldots, n-1\}^2$ which may be the empty set.

Let N be the cardinality of V^*. Then $\mathcal{H}(I)^N$ denotes the N-fold product of $(\mathcal{H}(I), h)$ equipped with the maximum metric h_∞, i.e. $h_\infty(\mathbf{A}, \mathbf{B}) = \max_{v \in V^*} h(A_v, B_v)$, where $\mathbf{A} = (A_v)_{v \in V^*}, \mathbf{B} = (B_v)_{v \in V^*} \in \mathcal{H}(I)^N$.

Definition 3.1 *Let $V^* = V \setminus \{v_0\}$ be a finite set of cardinality $N \geq 1$ and let $n \in \mathbb{N}$. A mapping $\mathcal{F} : \mathcal{H}(I)^N \to \mathcal{H}(I)^N$ is called n-adic-HIFS if and only if $\mathcal{F}$ satisfies the following properties:*

For each pair $(v, w) \in V^* \times V^*$ *there is a (possibly empty) set* $J(v, w) \subset \{0, \ldots, n-1\}^2$ *such that the* v*-component* ($v \in V^*$) $\mathcal{F}(\mathbf{A})_v$ *of* $\mathcal{F}(\mathbf{A})$ *has the form*

$$\mathcal{F}(\mathbf{A})_v = \bigcup_{w \in V^*} F_{vw}(A_w) \neq \emptyset \text{ for all } v \in V^*, \tag{4}$$

where

$$F_{vw} : \mathcal{H}(I) \cup \{\emptyset\} \to \mathcal{H}(I) \cup \{\emptyset\}$$
$$A \mapsto \bigcup_{(\alpha,\beta) \in J(v,w)} f_{(\alpha,\beta)}(A).$$

Remarks 3.1

1. The n-adic-HIFS given by $\mathcal{F}$ is a contraction on $(\mathcal{H}(I)^N, h_\infty)$ and thus has a unique attractive fixed point $\mathcal{A} = (A_v)_{v \in V^*}$, cf. [MAU88, EDG90], called the attractor of $\mathcal{F}$.

2. Each assignment $(v, w) \mapsto J(v, w)$ defines an n-adic-HIFS provided (4) is satisfied.

Let $\sigma : V \to \mathcal{M}_n$ be a matrix substitution system. Then we claim that σ defines an n-adic-HIFS by setting

$$J(v, w) = \{(\alpha, \beta) \mid \sigma(\underline{v})(\alpha, \beta) = w\}$$

where $(v, w) \in V^* \times V^*$ and $(\alpha, \beta) \in \{0, \ldots, n-1\}^2$. We have to show that this assignment $(v, w) \mapsto J(v, w)$ satisfies (4). Let us assume the contrary, i.e. for some $v \in V^*$

$$\mathcal{F}(\mathbf{A})_v = \bigcup_{w \in V^*} F_{vw}(A_w) = \bigcup_{w \in V^*} \bigcup_{(\alpha,\beta) \in J(v,w)} f_{(\alpha,\beta)}(A_w) = \emptyset.$$

Then

$$\bigcup_{w \in V^*} J(v, w) = \bigcup_{w \in V^*} \{(\alpha, \beta) \mid \sigma(\underline{v})(\alpha, \beta) = w\}$$

is empty, and this yields that $\sigma(\underline{v}) = \underline{v_0}$, which is a contradiction to the definition of a matrix substitution system.

We denote the n-adic-HIFS which is induced by a matrix substitution system σ by $\mathcal{F}_\sigma$.

Proposition 3.2 *Let* $\sigma : V \to \mathcal{M}_n$ *be a matrix substitution system and* $\mathcal{F}_\sigma$ *be the induced n-adic-HIFS with attractor* $\mathcal{A} = (A_v)_{v \in V^*}$. *Then*

$$A_v = \mathcal{A}(\underline{v})$$

for all $v \in V^*$.

Proof : Indeed, if $\mathbf{I} = (I, \ldots, I) \in \mathcal{H}(I)^N$, then we have for the k-th iteration of $\mathcal{F}_\sigma$ that

$$\mathcal{F}_\sigma^k(\mathbf{I})_v = \frac{1}{n^k} G\Sigma^k(\underline{v})$$

for all $k \in \mathbb{N}_0$ and all $v \in V^*$. □

Thus, matrix substitution systems can be characterized by n-adic-HIFS. But n-adic-HIFS lead also to MSSs. Indeed, let $\mathcal{F}$ be an n-adic-HIFS such that $J(v,w) \cap J(v',w') = \emptyset$ whenever $(v,w) \neq (v',w')$. Then $\mathcal{F}$ induces a matrix substitution system $\sigma : V \to \mathcal{M}_n$ by setting $\sigma(v_0) = \underline{v_0}$, and for $v \in V^*$ we set

$$\sigma(v)(\alpha,\beta) = \begin{cases} w & (\alpha,\beta) \in J(v,w) \\ v_0 & \text{otherwise} \end{cases}$$

for $(\alpha,\beta) \in \{0,\ldots,n-1\}^2$.

Remarks 3.2

1. Let $\mathcal{F}$ be an n-adic-HIFS and let σ be the induced matrix substitution system. Then $\mathcal{A}(\underline{v}) = A_v$ for all $v \in V^*$.

2. If σ is a matrix substitution system and $\mathcal{F}_\sigma$ the induced n-adic-HIFS, then $\mathcal{F}_\sigma$ satisfies *the open set condition*,see [MAU88].

3. If $\mathcal{F}$ is an n-adic-HIFS which satisfies the *open set condition*, then $\mathcal{F}$ induces a matrix substitution system σ. Indeed, the condition $J(v,w) \cap J(v',w') = \emptyset$, whenever $(v,w) \neq (v',w')$, is equivalent to the *open set condition*.

4 Linear cellular automata

The purpose of this section is to discuss the global pattern formation of linear cellular automata through attractors of n-adic-HIFS. We present an approach to linear cellular automata (LCA) which uses polynomials over the field $\mathbb{Z}_p$, where $p \in \mathbb{N}$ is a prime number. Other approaches and more details on LCA may be found e.g. in [HED69, MAR84, TAK90, WIL84, WI87a, WI87b].

Willson, see [WIL84], has associated a fractal set to a LCA using a proper scaling. This fractal set models in some sense the long term behavior of a LCA.

In the previous sections we introduced matrix substitution systems and n-adic-HIFS. Both methods generate fractal sets provided a suitable graphical representation is chosen.

The goal of this section is to show that Willson's fractal set associated to a LCA has an appropriate description as a limit set $\mathcal{A}(\omega)$ of a matrix substitution system, and for that matter as an attractor of an associated n-adic-HIFS. This gives a mathematical frame for some general claims and examples in [TAK90, TAK90]. Our result here is a special case of a result in [HAE91], where we will treat LCA in arbitrary dimensions.

From now on let $p \in \mathbb{N}$ be a prime number. By $\mathbb{Z}_p[x]$ we denote the ring of polynomials with coefficients in the field $\mathbb{Z}_p$.

Definition 4.1 *Let $r \in \mathbb{Z}_p[x]$ be a polynomial of degree $\deg r = d \geq 1$. Then r induces a linear mapping $\mathbf{r} : \mathbb{Z}_p[x] \to \mathbb{Z}_p[x]$ defined by $\mathbf{r}(s)(x) = r(x)s(x)$. The maping $\mathbf{r}$ is called a linear cellular automaton (with p states).*

Remarks 4.1

1. When we refer to a LCA $\mathbf{r}$ we allways will assume implicitely that there exists a generating polynomial $r \in \mathbb{Z}_p[x]$ which induces the linear mapping $\mathbf{r}$ on $\mathbb{Z}_p[x]$.

2. We will write $s(x) = \sum_{j=0}^{\infty} a_j x^j$, where $a_j = 0$ whenever $j > \deg s$, for a polynomial $s \in \mathbb{Z}_p[x]$.

3. We also introduce the following notation: Let $s \in \mathbb{Z}_p[x]$ be the polynomial $s(x) = \sum_{j=0}^{\infty} a_j x^j$, then $[s]_j := a_j$ and $[s]_j := 0$ for $j < 0$.

Moreover, we denote by $\mathbf{r}^k$ the k-th iterate of the mapping $\mathbf{r}$ (note that $\mathbf{r}^k$ is simply the multiplication with $r(x)^k = r^k$). As a direct consequence of Fermat's theorem (cf. [HAR79], p.63) we obtain the important relation

$$r(x)^p = r(x^p). \tag{5}$$

Definition 4.2 *The mapping $H : \mathbb{Z}_p[x] \to \mathcal{H}(\mathbb{R}^2) \cup \{\emptyset\}$ defined by*

$$H(s) = H(\sum_{j=0}^{\infty} a_j x^j) = \bigcup_{a_j \neq 0} I_{j,0}$$

is a graphical representation of polynomials in $\mathbb{Z}_p[x]$.

Remark 4.2

There is a strong analogy to the graphical representation of elements in $\mathcal{M}$ definned in section 2. Indeed, if $h : \mathbb{Z}_p^* \to \mathcal{H}(R^2)$ ($\mathbb{Z}_p^* = \mathbb{Z}_p \setminus \{0\}$) is any mapping, then h induces a graphical representation $\tilde{H}$ by

$$\tilde{H}(s) = \bigcup_{a_j \neq 0} h(a_j) + (j, 0).$$

In order to study the evolution of a LCA $\mathbf{r} : \mathbb{Z}_p[x] \to \mathbb{Z}_p[x]$ we have to consider the orbits $\{\mathbf{r}^k(s)\}_{k \in \mathbb{N}}$ for polynomials $s \in \mathbb{Z}_p[x], s \not\equiv 0$. The set

$$X(n, s) = \bigcup_{k=0}^{n-1} H(\mathbf{r}^k(s)) + (0, k)$$

is a graphical representation of the orbit of s (up to the $(n-1)$-st iterate). Let us now state Willson's scaling result.

Theorem 4.3 ([WIL84]) *Let* $\mathbf{r} : \mathbb{Z}_p[x] \to \mathbb{Z}_p[x]$ *be an LCA, and let* $s \in \mathbb{Z}_p[x], s \not\equiv 0$. *Then the sequence*

$$\left(\frac{1}{p^k} X(p^k, s)\right)_{k \in \mathbb{N}}$$

is a Cauchy sequence in $(\mathcal{H}(\mathbb{R}^2), h)$. *The limit of this sequence is independent of* s *and is denoted by* $Y(r)$.

Remarks 4.3

1. $Y(r)$ is called the limit set of the LCA $\mathbf{r}$.

2. $Y(r) \subset \cup_{j=0}^{d} I_{j,0}$, where $d = \deg r$.

3. As an immediate consequence of Fermat's theorem one has: If $I_{l,m} \subset X(p^k, 1)$ and $m \geq 1$ then $I_{(pl,pm)} \subset X(p^{k+1}, 1)$.

4. If $h : \mathbb{Z}^* \to \mathcal{H}(\mathbb{R}^2)$ is any mapping and $\tilde{H}$ is the induced graphical representation of polynomials, then

$$\tilde{X}(n, s) = \bigcup_{k=0}^{n-1} \tilde{H}(\mathbf{r}^k(s)) + (0, k)$$

is a graphical representation of the orbit of $s \in \mathbb{Z}_p[x]$, and

$$\lim_{k \to \infty} \frac{1}{p^k} \tilde{X}(p^k, s) = Y(r)$$

for all $s \in \mathbb{Z}_p[x], s \not\equiv 0$.

We now consider the orbit $\{r(x)^m\}_{m \in \mathbb{N}}$ of the polynomial $s \equiv 1$ under the iteration of $\mathbf{r}$. We write

$$r(x)^m = \sum_{l=0}^{\infty} a_l(m) x^l, m \in \mathbb{N},$$

where $a_l(m) = 0$ for all $l > m \deg(r)$. Due to Fermat's theorem we have $a_l(m) = a_{pl}(pm)$ and $a_{pl+j}(pm) = 0$ for $j = 1, \ldots, p-1$. This yields the following scheme

$$\begin{array}{cc} a_l(m+1) & a_{l+1}(m+1) \\ a_l(m) & a_{l+1}(m) \end{array} \quad \Bigg| \quad \begin{array}{ccc} a_{pl}(p(m+1)) & 0\ldots0 & a_{p(l+1)}(p(m+1)) \\ \vdots & \cdots & \vdots \\ a_{pl}(pm) & 0\ldots0 & a_{p(l+1))}(pm). \end{array}$$

The graphical representation of the orbit $\{r(x)^m\}_{m=0}^{p^k-1}$ for k suffiently large, and the scaling show that the square $I_{l,m}$ is replaced by the squares

$$\bigcup_{i,j \in \{0,\ldots,p-1\}} I_{pl+i,pm+j}.$$

It therefore remains to determine the coefficients $a_{pl+i}(pm+j)$, where $i, j \in \{0, \ldots, p-1\}$. Since the polynomial r is of degree d, the coefficient $a_l(m+1)$ is determined by the coefficients $a_{l-d+1}(m), a_{l-d+2}(m), \ldots, a_l(m)$, where we set $a_j(m) = 0$ if $j < 0$. We thus obtain

$$\begin{array}{c|ccc} & a_{pl}(p(m+1)) & 0\ldots0 & a_{p(l+1)}(p(m+1)) \\ a_{pl-d+1}(p(m+1)-1)\ldots & a_{pl}(p(m+1)-1) & \ldots & a_{p(l+1)}(p(m+1)-1) \\ \vdots & \vdots & \vdots & \vdots \\ \ldots & a_{pl}(pm) & 0\ldots0 & a_{p(l+1)}(pm). \end{array}$$

This shows that all coefficients $a_{pl+i}(pm+j)$, where $i, j \in \{0, \ldots, p-1\}$, together with their $d-1$ left neighbours are determined by $a_{l-d+1}(m), a_{l-d+2}(m), \ldots, a_l(m)$.

In terms of polynomials we may formulate our observation in the following way. For the coefficients $a_{l-d+1}(m), a_{l-d+2}(m), \ldots, a_l(m)$ we consider the polynomial $s(x) = a_{l-d+1}(m) + a_{l-d+2}(m)x + \ldots + a_l(m)x^{d-1}$. Then the coefficient $a_{pl+i}(pm+j)$ is given by $[s(x^p)r(x)^j]_{p(d-1)+i}$.

We are now prepared to state the main result, i.e. we are able to define a matrix substitution system which models a given LCA. But first we need an appropriate set of symbols: Let $V = \{(b_{d-1}, b_{d-2}, \ldots, b_0) \,|\, b_j \in \mathbb{Z}_p, j = 0, 1, \ldots, d-1\}$ be our set of symbols with distinguished element $(0, \ldots, 0)$. We then define a substitution $\sigma_r : V \to \mathcal{M}_p$. Let $v = (b_{d-1}, \ldots, b_0) \in V$, then we set

$$\sigma_r(v)(\alpha, \beta) = (w(\alpha, \beta)_j)_{j=d-1}^{0} \in V$$
$$\text{where } w(\alpha, \beta)_j = [s_v(x^p)r(x)^\beta]_{p(d-1)+\alpha-j}$$

and $s_v(x) = b_{d-1} + b_{d-2}x + \ldots + b_0x^{d-1}$ and $\alpha, \beta \in \mathbb{Z}_p$.

Remark 4.4

Using the linearity of the above construction it sufficies to calculate $\sigma_r(e_j)$ for $j = 0, 1, \ldots, d-1$, where $e_j = ((\epsilon_j)_k)$ and $(\epsilon_j)_k = 0$ if $k \neq j$, and $(\epsilon_j)_k = 1$ if $k = j$.

Theorem 4.4 *Let* $\mathbf{r} : \mathbb{Z}_p[x] \to \mathbb{Z}_p[x]$ *be a LCA of degree* d*. Then the mapping* $\sigma_r : V \to \mathcal{M}_p$ *as defined above is a matrix substitution system. Furthermore*

$$\bigcup_{j=0}^{d-1} \mathcal{A}(\underline{e_j}) + (j, 0) = Y(r).$$

Proof : By construction of σ_r it is obvious that σ_r is a matrix substitution system. Since $X(p^k, 1) \subset G\Sigma^k(e_0e_1 \ldots e_{d-1})$ and $h(X(p^k, 1), G\Sigma^k(e_0e_1 \ldots e_{d-1})) \leq d$ the assertion follows. □

Remark 4.5

Using a special graphical representation of elements in V it is possible to model the evolution of the LCA exactly by the $\mathcal{M}_p$-substitution σ_r.

Indeed, for $v = (b_j)_{j=d-1}^{0}$ just define a graphical representation of elements in V by

$$g(v) = \bigcup_{j=0}^{d-1} \eta(b_j) - (j, 0)$$

where $\eta : \mathbb{Z}_p \to \mathcal{H}(\mathbb{R}^2) \cup \{\emptyset\}$ is defined by

$$\eta(b_j) = \begin{cases} \emptyset & \text{if } b_j = 0 \\ I_{0,0} & \text{if } b_j \neq 0 \end{cases} .$$

Then the induced graphical representation $\tilde{G}$ has the property

$$\tilde{G}\Sigma_r^k(e_0 e_1 \ldots e_{d-1}) = X(p^k, 1)$$

for all $k \in \mathbb{N}$.

5 Dimension

In this section we calculate the box-counting and Hausdorff dimension of limit sets of matrix substitution systems, see [FAL85, EDG90] for definitions, and as a consequence of the previous sections, we will be able to calculate the dimension of the attractor of an n-adic-HIFS and the dimension of the limit set of a LCA as well.

We start with the calculation of the box-counting dimension of the limit set of a matrix substitution system.

Let $\sigma : V \to \mathcal{M}_n$ be a matrix substitution system with induced mapping $\Sigma : \mathcal{M} \to \mathcal{M}$. We choose a linear order in V with first element v_0. The set $V \setminus \{v_0\}$ equipped with this linear order is denoted by V'. Let $N \geq 1$ denote the cardinality of V', and let $\mathbb{Z}^N$ be the standard integer lattice in the N-dimensional Euclidian space $\mathbb{R}^N$. Coordinates of points in $\mathbb{R}^N$ will be indexed by elements of V'. We need a few definitions.

Definition 5.1 *The mapping $b : \mathcal{M} \to \mathbb{R}^N$ defined by*

$$(b(\omega)_v)_{v \in V'} = (card\{(l, m) \mid (l, m) \in \mathbb{N}_0^2 \text{ such that } \omega(l, m) = v\}_v)_{v \in V'}$$

is called symbol counting map.

Remarks 5.1

1. The v-component of $b(\omega)$ equals the number of symbols v in the matrix ω.

2. The number $N(\omega)$ of all non-trivial symbols of the matrix $\omega \in \mathcal{M}_n$ is given by

$$N(\omega) = \sum_{v \in V'} b(\omega)_v ,$$

or $N(\omega) = (e|b(\omega))$, where $(.|.) : \mathbb{R}^N \times \mathbb{R}^N \to \mathbb{R}$ denotes the standard scalar product on $\mathbb{R}^N$ and $e = (1, \ldots, 1)$.

Definition 5.2 *The matrix* $T = (t_{v,w})_{(v,w)\in V'\times V'}$ *with entries*

$$t_{v,w} = card\{(\alpha,\beta)\,|\,(\alpha,\beta)\in\{0,\dots,n-1\}^2 \text{ such that } \sigma(v)(\alpha,\beta)=w\}$$

is called transition matrix (of the matrix substitution system σ*).*

Remarks 5.2

1. For all $(v,w)\in V'\times V'$ one has $0\le t_{v,w}\le n^2$.

2. The set

$$R(v) = \{w\in V'\,|\,\text{there are } k\in\mathbb{N}_0, (l,m)\in\mathbb{N}_0^2 \text{ such that } \Sigma^k(\underline{v})(l,k)=w\}$$

   is called the range of $v\in V'$.

3. The range $R(v)$, $v\in V'$, defines a submatrix $T(v)$ of T which is

$$T(v) = (t_{w_1,w_2})_{w_1,w_2\in R(v)}.$$

   $T(v)$ is called the transition matrix (of σ) with respect to v.

Lemma 5.3 *The number of non-trivial symbols in the matrix* $\Sigma^k(\underline{v})$, $v\in V'$ *is given by*

$$N(v,k) = (e|T^k b(\underline{v})).$$

If $A\in\mathbb{R}^{m\times m}$ is any matrix with non-negative elements, then $\lambda_+(A)$ denotes the dominant eigenvalue of A ([GAN59]).

Proposition 5.4 *The box-counting dimension* $D(\underline{v})$ *of the limit set* $\mathcal{A}(\underline{v})$, $v\in V'$, *is given by*

$$D(\underline{v}) = \log_n \lambda_+(T(v)).$$

Proof : The arguments given in [WI87b], p. 197-199, apply. □

Theorem 5.5 *Let* $\mathbf{r}:\mathbb{Z}_p[x]\to\mathbb{Z}_p[x]$ *be a LCA of degree* d *and* σ_r *be the induced matrix substitution system. The box-counting dimension* D *of* $Y(r)$ *is*

$$D = \max\{\log_p \lambda_+(T(e_j))\,|\,j=0,\dots,d-1\}$$

where the e_j *are as in remark 4.4.*

We now turn our attention to the calculation of the Hausdorff dimension of the attractor of a n-adic-HIFS. It is a well known fact that for an n-adic-HIFS the box-counting dimension and the Hausdorff dimension coincide (see [FAL85], pp. 118-120, [EDG90], p. 184). The same result follows from [MAU88]. Thus from theorem 5.5 we obtain

Corollary 5.6 ([WI87a]) *The box-counting and Hausdorff dimension of the limit set* $Y(r)$ *of a LCA* $\mathbf{r}$ *coincide.*

A matrix substitution system is called mixing if $R(v) = V'$ for all $v \in V'$. In this case the Hausdorff dimension of the limit sets $\mathcal{A}(\underline{v})$ is equal $\lambda_+(T)$.

6 Examples

In our final section we offer some examples which are chosen to demonstrate the ideas presented in the theoretical sections. We start with a polynomial $r \in \mathbb{Z}_p[x]$ for $p = 2$, then we state the associated matrix substitution system and finally we calculate the Hausdorff dimension of the limit set. The computer graphical results in this section were produced with the software [JUE91].

1. Let $r(x) = 1 + x$, then $\deg(r) = 1$. $V = \{0, 1\}$ and define $\sigma : V \to \mathcal{M}_2$ by

$$\sigma(0) = \begin{vmatrix} 0 & 0 \\ 0 & 0 \end{vmatrix} \quad \sigma(1) = \begin{vmatrix} 1 & 1 \\ 1 & 0 \end{vmatrix}$$

The transition matrix is given by $T = |3|$. Therefore the Hausdorff dimension of the attractor is $\log_2 3$.

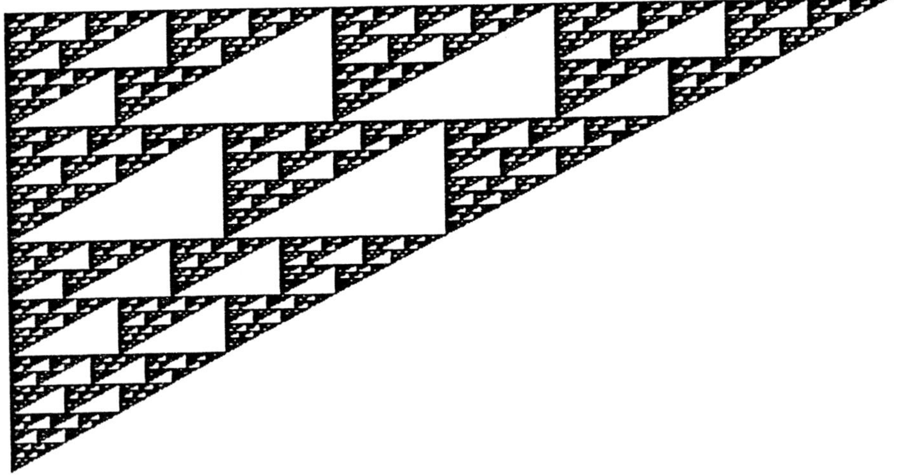

Figure 2 : The attractor of the LCA $r(x) = 1 + x + x^2$

2. Let $r(x) = 1 + x + x^2$. Set $V = \{00, 01, 10, 11\}$ and define σ by

$$\sigma(00) = \begin{vmatrix} 00 & 00 \\ 00 & 00 \end{vmatrix} \quad \sigma(01) = \begin{vmatrix} 01 & 11 \\ 01 & 10 \end{vmatrix}$$

$$\sigma(10) = \begin{vmatrix} 11 & 10 \\ 00 & 00 \end{vmatrix} \quad \sigma(11) = \begin{vmatrix} 10 & 01 \\ 01 & 10 \end{vmatrix}$$

The transition matrix is given by

$$T = \begin{vmatrix} 2 & 0 & 2 \\ 1 & 1 & 2 \\ 1 & 1 & 0 \end{vmatrix}$$

which has the dominant eigenvalue $\lambda_+(T) = 1 + \sqrt{5}$. Therefore the Hausdorff dimension of the attractor is $\log_2(1 + \sqrt{5})$, see [WIL84, TAK90].
3. Let $r(x) = 1 + x + x^3$. Then $V = \{000, 001, 010, 011, 100, 101, 110, 111\}$ and define σ by

$$\sigma(000) = \begin{vmatrix} 000 & 000 \\ 000 & 000 \end{vmatrix} \quad \sigma(001) = \begin{vmatrix} 001 & 011 \\ 001 & 010 \end{vmatrix} \quad \sigma(010) = \begin{vmatrix} 110 & 101 \\ 100 & 000 \end{vmatrix}$$

$$\sigma(011) = \begin{vmatrix} 111 & 110 \\ 101 & 010 \end{vmatrix} \quad \sigma(100) = \begin{vmatrix} 010 & 100 \\ 000 & 000 \end{vmatrix} \quad \sigma(101) = \begin{vmatrix} 011 & 111 \\ 001 & 010 \end{vmatrix}$$

$$\sigma(110) = \begin{vmatrix} 100 & 001 \\ 100 & 000 \end{vmatrix} \quad \sigma(111) = \begin{vmatrix} 101 & 010 \\ 101 & 010 \end{vmatrix}$$

The transition matrix is given by

$$T = \begin{vmatrix} 2 & 0 & 0 & 0 & 1 & 1 & 0 \\ 1 & 0 & 1 & 1 & 1 & 0 & 2 \\ 1 & 0 & 0 & 0 & 1 & 0 & 0 \\ 0 & 1 & 0 & 1 & 0 & 2 & 0 \\ 0 & 1 & 1 & 0 & 0 & 0 & 2 \\ 0 & 1 & 1 & 0 & 0 & 0 & 0 \\ 0 & 0 & 1 & 0 & 1 & 0 & 0 \end{vmatrix}$$

and $\lambda_+(T) \approx 3.311422085\ldots$.

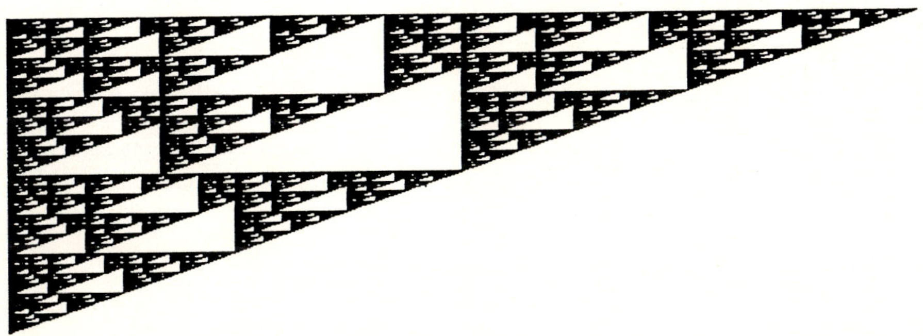

Figure 3 : The attractor of the LCA $r(x) = 1 + x + x^3$

4. Sometimes it is useful to generalize the notion of a linear cellular automaton. Let $\mathbb{Z}_p(x)$ denote the set of finite Laurent series with coefficients in $\mathbb{Z}_p$, i.e. $s \in \mathbb{Z}_p(x)$ if and only if $s(x) = \sum_{j=-\infty}^{\infty} a_j x^j$ and there exists a $K \in \mathbb{N}$ such that $a_j = 0$ for all $j \in \mathbb{Z}$ with $|j| > K$. A finite Laurent series $r \in \mathbb{Z}_p(x)$ induces a linear mapping $\mathbf{r} : \mathbb{Z}_p(x) \to \mathbb{Z}_p(x)$ by $\mathbf{r}(s)(x) = r(x)s(x)$. Again, the mapping $\mathbf{r}$ is called a linear cellular automaton. A straightforward calculation shows that the results of section 4 can be extended to LCA defined by a finite Laurent series.

Let $r(x) = x^{-1} + 1 + x \in \mathbb{Z}_2(x)$ and consider the set of symbols $V = \{00, 01, 10, 11\}$. We define $\sigma : V \to \mathcal{M}_2$ by

$$\sigma(00) = \begin{vmatrix} 00 & 00 \\ 00 & 00 \end{vmatrix} \quad \sigma(01) = \begin{vmatrix} 01 & 11 \\ 00 & 10 \end{vmatrix}$$

$$\sigma(10) = \begin{vmatrix} 11 & 10 \\ 10 & 00 \end{vmatrix} \quad \sigma(11) = \begin{vmatrix} 10 & 01 \\ 10 & 01 \end{vmatrix}$$

Then the limit set $Y(r)$ of the LCA $\mathbf{r}$ is given by $Y(r) = (\mathcal{A}(\underline{01}) + (-1, 0)) \cup (\mathcal{A}(\underline{10}) + (0, 0))$.

Figure 4 : The limit set $Y(r)$ of $r(x) = x^{-1} + 1 + x$

5. Using the invariance property $r(x^{-1}) = r(x)$ of the above Laurent series we can simplify the n-adic-HIFS induced by σ to an ordinary IFS defined on $\mathcal{H}(I)$. We present a symbolic notation for this IFS

<table>
<tr><td rowspan="2" colspan="2">$\bar{I}$</td><td>$\bar{I}$</td><td>I</td></tr>
<tr><td>$\bar{I}$</td><td>I</td></tr>
<tr><td colspan="2" rowspan="2"></td><td colspan="2" rowspan="2">I</td></tr>
<tr></tr>
</table>

Here I denotes a scaled image of the unit square, and $\overline{I}$ is a scaled image of the reflection of the unit square at the vertical line $\{(x, y) \in \mathbb{R}^2 \,|\, x = 1/2\}$. The scaling factors are obvious from the picture. The attractor of the IFS is $\mathcal{A}(\underline{0}\underline{1})$, see [TAK90]. This example proposes the general problem to study how n-adic-HIFS which are induced by LCA can be reduced, see also the following example.

6. Let $r(x) = x^{-2} + x^{-1} + 1 + x + x^2 \in \mathbb{Z}_2(x)$. The set of symbols is given as $V = \{b_3b_2b_1b_0 \,|\, b_j \in \mathbb{Z}_2, j = 1,2,3,4\}$. To simplify the notation we identify a symbol $b_3b_2b_1b_0$ with the number $b_0 + 2b_1 + 4b_2 + 8b_3$. Then the matrix substitution system induced by $\mathbf{r}$ is given by

$$\sigma(0) = \begin{vmatrix} 0 & 0 \\ 0 & 0 \end{vmatrix} \quad \sigma(1) = \begin{vmatrix} 1 & 3 \\ 0 & 0 \end{vmatrix} \quad \sigma(2) = \begin{vmatrix} 7 & 15 \\ 1 & 2 \end{vmatrix} \quad \sigma(3) = \begin{vmatrix} 6 & 12 \\ 1 & 2 \end{vmatrix}$$

$$\sigma(4) = \begin{vmatrix} 15 & 14 \\ 4 & 8 \end{vmatrix} \quad \sigma(5) = \begin{vmatrix} 14 & 13 \\ 4 & 8 \end{vmatrix} \quad \sigma(6) = \begin{vmatrix} 8 & 1 \\ 5 & 10 \end{vmatrix} \quad \sigma(7) = \begin{vmatrix} 9 & 2 \\ 5 & 10 \end{vmatrix}$$

$$\sigma(8) = \begin{vmatrix} 12 & 8 \\ 0 & 0 \end{vmatrix} \quad \sigma(9) = \begin{vmatrix} 13 & 11 \\ 0 & 0 \end{vmatrix} \quad \sigma(10) = \begin{vmatrix} 11 & 7 \\ 1 & 2 \end{vmatrix} \quad \sigma(11) = \begin{vmatrix} 10 & 4 \\ 1 & 2 \end{vmatrix}$$

$$\sigma(12) = \begin{vmatrix} 3 & 6 \\ 4 & 8 \end{vmatrix} \quad \sigma(13) = \begin{vmatrix} 2 & 5 \\ 4 & 8 \end{vmatrix} \quad \sigma(14) = \begin{vmatrix} 4 & 9 \\ 5 & 10 \end{vmatrix} \quad \sigma(15) = \begin{vmatrix} 5 & 10 \\ 5 & 10 \end{vmatrix}$$

The induced n-adic-HIFS is defined on $\mathcal{H}(I)^{15}$. Using the invariance property of r we are able to simplify the induced n-adic-HIFS. We thus obtain a hierarchical iterated function system defined on $\mathcal{H}(I)^4$ and which we denote symbolically as

$I_1 =$

I_1	I_3

$I_3 =$

$\overline{I_1}$	I_1	$\overline{I_3}$
I_5	$\overline{I_5}$	
I_1		I_2

$I_2 =$

C	I_2	I_5	$\overline{I_5}$
I_5	$\overline{I_5}$	I_5	$\overline{I_5}$
I_1		I_2	

$I_5 =$

$\overline{I_2}$	C	I_2	I_5
I_5	$\overline{I_5}$	$\overline{I_2}$	$\overline{I_1}$
$\overline{I_2}$		$\overline{I_1}$	

where C is defined as

$C =$

I_2	I_5	$\overline{I_5}$	$\overline{I_2}$
$\overline{I_2}$	$\overline{I_1}$	I_1	I_2

Figure 5 shows the 5th iterate of the above defined HIFS. Shown are the spaces I_1 and I_2, which coresspond to the left part of the symmetric attractor.

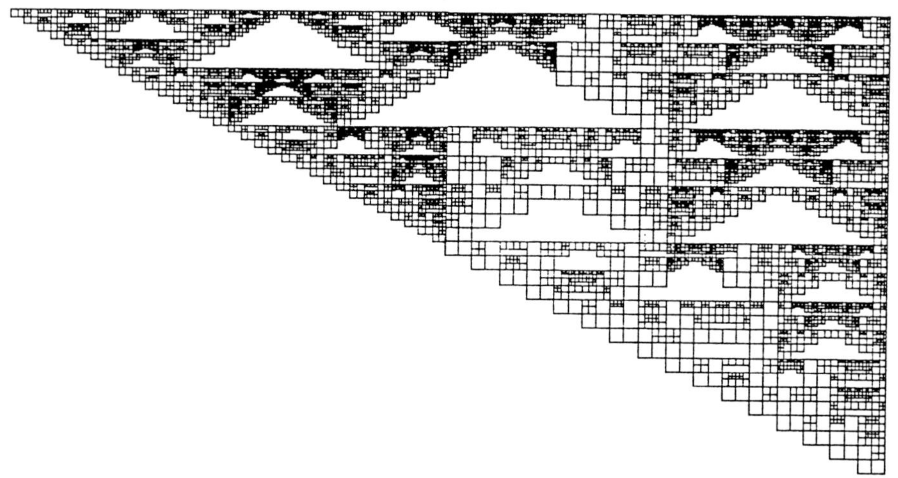

Figure 5 : The 5th iterate of the HIFS asociated to $r(x) = x^{-2} + x^{-1} + 1 + x + x^2$

Figure 6 : The left part of $Y(r)$ for $r(x) = x^{-2} + x^{-1} + 1 + x + x^2$

7 References

[BAN] C. Bandt: Self-similar sets 3. Constructions with sofic systems, Monatshefte Math. 108 (1989), 89-102

[BAR85] M. Barnsley, S. Demko: Iterated function systems and the global construction of fractals, Proc. Roy. Soc., London, A399(1985), 243

[BAR88] M. Barnsley: Fractals everywhere, Acad. Press, Inc., San Diego, 1988

[BA88a] M. Barnsley, M. Berger, H. Soner: Mixing Markov chains and their images, Prob. Eng. Inf. Sci., 2(1988), 387-414

[BAR89] M. Barnsley, J. Elton, D. Hardin: Reccurent iterated function systems, Constr.Appr., 5(1989), 3-32

[BER88] M. Berger: Encoding images through transition probabilities, Math. Comput. Modelling, 11, (1988), 575-577

[BER89] M. Berger: Images generated by orbits of 2-D Markov chains (1989), Chance: New Directions for Statistics and Computing Vol. 2, No. 2, 18-28

[BED86] T. Bedford: Dynamics and dimension for fractal recurrent sets, J. London Math. Soc. (2), 33 (1986), 89-100

[DEK82] F.M. Dekking: Recurrent Sets, Advances in Mathematics 44, 1 (1982), 78-104

[DEK89] F.M. Dekking: Substitutions, branching processes and fractal sets, to appear in Proc. NATO ASI on Fractal Geometry, Montreal, July 1989, Kluwer Acad. Publ.

[DEK90] F.M. Dekking, R.W.J. Meester: On the structure of Mandelbrot's percolation process and other random Cantor sets, J. Statist. Physics, 58, 5/6 (1990), 1109-1126

[EDG90] G. Edgar: Measure, Topology and Fractal Geometry, Springer Verlag, New York,1990

[FAL85] K.J. Falconer: The Geometry of Fractal sets, Cambridge Univ. Press, N.Y., 1985

[GAN59] F.R. Gantmacher: The Theory of Matrices, vol. 2,Chelsea, New York, 1959

[HAE91] F. v.Haeseler, H.-O. Peitgen, G. Skordev: Pascal's triangle, dynamical systems, and attractors, submitted.

[HAE91] F. v.Haeseler, H.-O. Peitgen, G. Skordev: Linear cellular automata and HIFS, in preparation

[HAR79] G.H. Hardy, E.M. Wright: An Introduction to the Theory of Numbers, Oxford at the Clarendon Press, 1979, 5th edition

[HED69] G. Hedlung: Endomorphisms and automorphisms of the shift dynamical systems, Math. Syst. Th. 3, 4 (1969), 320-375

[HUT81] J.E. Hutchinson: Fractals and selfsimilarity, Indiana Math. J., 30 (1981), 713-747

[JUE91] H. Jürgens, H.-O. Peitgen, D. Saupe: IFS-Tutor, Software Package for Hierarchical Iterated Function Systems, 1991, Universität Bremen

[KUM52] E.E. Kummer: Über die Ergänzungssätze zu den allgemeinen Reciprocitätsgesetzen, J. reine. angew. Math. 44(1852), 93-146.

[MAN82] B. Mandelbrot: The Fractal Geometry of Nature, W.H. Freeman and Co., New York, 1982

[MAN85] B. Mandelbrot, Y. Gefen, A. Aharony, J. Peyriere: Fractals, their transfer matrices and their eigen-dimensional sequences, J. Phys. A: Math. Gen., 18 (1985), 335-354

[MAR84] O. Martin, A. Odlyzko, S. Wolfram: Algebraic properties of cellular automata, Commun. Math. Physics, 93 (1984), 219-258

[MAU88] H.D. Mauldin, S.C. Williams: Hausdorff dimension in graph directed constructions, Trans. Amer. Math. Soc., 309, 2 (1988),811-829

[PRU90] P. Prusinkiewicz, A. Lindenmayer: The Algorithmic Beauty of Plants, Springer Verlag, New York, 1990

[TAK90] S. Takahashi: Cellular automata and multifractals: dimension spectra of linear cellular automata, Physica D 45 (1990), 36-48

[TAK91] S. Takahashi: Self-similarity of linear cellular automata, J. Comp. Sci. (to appear)

[WIL71] R. Williams: Composition of contractions, Bol. Soc. Brasil. Mat., 2 (1971), 55-59

[WIL84] S. Willson: Cellular automata can generate fractals, Discrete Appl. Math. 8 (1984),91-99

[WI87a] S. Willson: The equality of fractional dimension for certain cellular automata, Physica 24D (1987), 179-189

[WI87b] S. Willson: Computing fractal dimensions for additive cellular automata, Physica 24D (1987), 190-206

[WOL83] S. Wolfram: Statistical mechanics and cellular automata, Rev. Mod. Phys. 55 (1983), 601-644

[WOL84] S. Wolfram: Universality and complexity in cellular automata, Physica 10D (1984), 1-35

Escape-time Visualization Method for Language-restricted Iterated Function Systems

P. Prusinkiewicz and M. S. Hammel
University of Calgary, Canada

Abstract: The escape-time method was introduced to generate images of Julia and Mandelbrot sets, then applied to visualize attractors of iterated function systems. This paper extends it further to language-restricted iterated function systems (LRIFS's). They generalize the original definition of IFS's by providing means for restricting the sequences of applicable transformations. The resulting attractors include sets that cannot be generated using ordinary IFS's. The concepts of this paper are expressed using the terminology of formal languages and finite automata.
Keywords: fractal, iterated function system, escape-time method, graphics algorithm, formal language, finite automaton.

1 Introduction

Although mathematicians have explored the properties of fractals since the turn of century, they could not visualize the objects of their study without the aid of computers. Computer graphics made it possible to recognize the beauty of fractals, and turned them into an an art form [10]. Peitgen and Richter [11] perfectioned and popularized images of Julia and Mandelbrot sets. Many of them were created using the escape-time method. In its original setting, it consisted of testing how fast points z outside the attractor diverged to infinity while iterating function $z \rightarrow z^2 + c$ in the complex plane. The resulting values were interpreted as colors in a two-dimensional image, or height values in a "fractal landscape" [12, Section 2.7].

The escape-time visualization method was extended from Julia sets to iterated function systems by Barnsley [2] and Prusinkiewicz and Sandness [15]. This paper extends it further to language-restricted iterated function systems [13]. They generalize the original definition of IFS's by providing means for restricting the sequences of applicable transformations to a particular set. The resulting attractors form a larger class than those generated using ordinary IFS's. The definition of an LRIFS leaves open the mechanism for sequencing transformations, thus LRIFS's incorporate the earlier generalizations committed to a particular mechanism, such as sofic systems [1], recurrent IFS's [3], Markov IFS's [18], mixed IFS's [4], controlled IFS's [14], and mutually recursive function systems [6, 7]. Several other authors considered similar generalizations without giving them a name.

The paper is organized as follows. Sections 2 and 3 summarize the background material related to formal languages and iterated function systems. Section 4 presents the escape-time method for IFS's in a way suitable for further extensions. Section 5 defines the language-restricted iterated function systems. The escape-time method is extended to LRIFS's in Section 6. A special case of regular languages is considered and illustrated using examples in Section 7. Section 8 summarizes the results.

2 Formal Languages

An *alphabet* V is as a finite nonempty set of *symbols* or *letters.* A *string* or *word* over alphabet V is a finite sequence of zero or more letters of V, whereby the same letter may occur several times. The total number of letters in a word w is called its *length*, and denoted length(w). The word of zero length is called the *empty word*, and denoted ϵ. The *concatenation* of words $x = a_1a_2\dots a_m$ and $y = b_1b_2\dots b_n$ is the word formed by extending the sequence of symbols x with the sequence y, thus $xy = a_1a_2\dots a_mb_1b_2\dots b_n$. If $xy = w$ then the word x is called the *prefix* of w, denoted $x \prec w$. The remaining word y is called the *suffix*. We assume that the relation $\prec$ is reflexive, that is $w \prec w$. The n-fold concatenation of a word w with itself is called its n-th power, and denoted w^n. By definition, $w^0 = \epsilon$ for any w. If $w = a_1a_2\dots a_n$, than the word $w^R = a_n\dots a_2a_1$ is called the *mirror image* of w. It can be shown that $(xy)^R = y^Rx^R$ for any words x and y.

The set of all words over V is denoted by $V^\star$, and the set of nonempty words by V^+. A *formal language* over an alphabet V is a set L of words over V, hence $L \subset V^\star$. The concatenation and mirror image of words are extended to languages as follows:

$$\begin{aligned} L_1L_2 &= \{xy : x \in L_1 \;\&\; y \in L_2\}, \\ L^R &= \{w^R : w \in L\}. \end{aligned}$$

A language L is *prefix extensible*, if there exists a word $v \in V^+$ such that $vL \subset L$. In other words, $vw \in L$ for every word $w \in L$. The *right derivative* of a language $L \subset V^\star$ with respect to a word $v \in V^\star$ is the language:

$$L//v = \{w \in V^\star : vw \in L\}.$$

The set of all prefixes of a language L is called the *prefix closure* of L:

$$\mathcal{P}(L) = \{x \in V^\star : (\exists w \in L)\, x \prec w\}.$$

3 Iteraded Function Systems

Let $\langle X, d\rangle$ be a complete metric space with support X and distance function d (in this paper, we will only consider the plane with the Euclidean distance). A function $F : X \to X$ is called a *contraction* in X if there is a constant $r < 1$ such that

$$d(F(P), F(Q)) \leq rd(P, Q)$$

for all $P, Q \in X$. The parameter r is called the *Lipschitz constant* of F.

An *iterated function system* (IFS) in X is a quadruplet $\mathcal{I} = \langle X, \mathcal{F}, V, h, \rangle$, where:

- X is the underlying metric space,
- $\mathcal{F}$ is a set of contractions in X,
- V is an alphabet of contraction labels,
- $h : V \to \mathcal{F}$ is a *labeling function*, taking the letters of alphabet V to the contractions from $\mathcal{F}$.

The labeling function h is extended to words and languages over V using the equations:

$$\begin{aligned} h(a_1 a_2 \ldots a_n) &= h(a_1) \circ h(a_2) \circ \ldots \circ h(a_n), \\ h(L) &= \bigcup_{w \in L} h(w), \end{aligned}$$

where $\circ$ denotes function composition,

$$x \circ f_1 \circ f_2 \circ \cdots \circ f_n = f_n(\cdots(f_2(f_1(x)))\cdots).$$

The *attractor* of an IFS $\mathcal{I}$ is the smallest nonempty set $\mathcal{A} \subset X$, closed with respect to all transformations of $\mathcal{F}$. Hutchinson showed that the attractor of an arbitrary IFS always exists and is unique [9]. Consequently, it can be found by selecting a point $P \in \mathcal{A}$, and applying to it all possible sequences of transformations from $\mathcal{F}$:

$$\mathcal{A} = P \circ h(V^\star).$$

There are several methods for finding the initial point $P \in \mathcal{A}$. For example, the fixed point of any transformation $F \in \mathcal{F}$ belongs to $\mathcal{A}$ [9].

A legible notation for specifying transformations is needed while defining particular IFS's. In this paper we express transformations by composing operations of translation, rotation, and scaling in an underlying Cartesian coordinate system. The following symbols are used:

- $t(a, b)$ is a translation by vector (a, b).
- $a(\alpha)$ is a rotation by (oriented) angle α with respect to the origin of the coordinate system. The angles are expressed in degrees.

Figure 1: The dragon curve

- $s(r_x, r_y)$ is a scaling with respect to the origin of the coordinate system: $x' = r_x x$ and $y' = r_y y$. If $r_x = r_y = r$, we write $s(r)$ instead of $s(r, r)$.

For example, Figure 1 shows the attractor of an IFS $\mathcal{I} = \langle X, \mathcal{F}, V, h \rangle$, where the set $\mathcal{F}$ consists of two transformations:

$$\begin{aligned} F_1 &= s(\frac{\sqrt{2}}{2}) \circ r(45), \\ F_2 &= s(\frac{\sqrt{2}}{2}) \circ r(135) \circ t(0,1). \end{aligned}$$

4 The Escape-time Method

Consider an IFS $\mathcal{I} = \langle X, \mathcal{F}, V, h \rangle$, where all functions $F \in \mathcal{F}$ are invertible. Let $\tilde{h}(a)$ denote the inverse of the contraction $F = h(a) \in \mathcal{F}$, thus $\tilde{h}(a) = (h(a))^{-1}$. The function $\tilde{h}$ is extended to words and languages over V in a way similar to h:

$$\begin{aligned} \tilde{h}(a_1 a_2 \ldots a_n) &= \tilde{h}(a_1) \circ \tilde{h}(a_2) \circ \ldots \circ \tilde{h}(a_n), \\ \tilde{h}(L) &= \bigcup_{w \in L} \tilde{h}(w). \end{aligned}$$

An *escape trajectory* of a point Q with respect to a word $w \in V^\star$ is the set

$$\mathrm{Tr}(Q, w) = \{Q \circ \tilde{h}(x) : x \prec w\}.$$

The length of w is referred to as the length of the trajectory. The escape-time method is based on the following Theorem, proven in [15]:

Theorem 1. (a) If a starting point Q belongs to the attractor $\mathcal{A}$ of an IFS $\mathcal{I}$, there exists an infinitely long trajectory entirely included in $\mathcal{A}$. (b) If the point Q does not belong to $\mathcal{A}$, all trajectories diverge to infinity.

To estimate the speed with which the divergence occurs, we enclose the attractor in a circle. Since attractors of IFS's are bounded, it is always possible to find a circle $\mathcal{C}$ of a finite radius R, completely enclosing $\mathcal{A}$. The escape time of a point $Q \notin \mathcal{A}$ is then defined as the length of the longest trajectory included in $\mathcal{C}$:

$$E_1(Q) = \max_{w \in V^*} \{\text{length}(w) : \text{Tr}(Q, w) \subset \mathcal{C}\}.$$

According to this definition, the function $E_1(Q)$ is integer-valued. In order to represent the escape time with a higher precision, Hepting *et al.* [8] introduced a residual term that reflects the distance between the last point in the escape trajectory $Tr(Q, w)$ and the border of circle $\mathcal{C}$:

$$E_2(Q) = \max_{wa \in V^*} \{\text{length}(w) + \text{res}(Q \circ \tilde{h}(w), a) : \text{Tr}(Q, w) \subset \mathcal{C}\}.$$

Let $Z = Q \circ \tilde{h}(w)$. The function res : $X \times V \rightarrow [0, 1)$ is defined as follows:

$$\text{res}(Z, a) = \begin{cases} \dfrac{\log R - \log \|Z\|}{\log \|Z \circ \tilde{h}(a)\| - \log \|Z\|} & \text{if } Z \in \mathcal{C} \text{ and } Z \circ \tilde{h}(a) \notin C, \\ 0 & \text{otherwise.} \end{cases} \tag{1}$$

The norm symbol $\|Z\|$ denotes the distance between point Z and the center O of circle $\mathcal{C}$, thus $\|Z\| = d(Z, O)$. An illustration of formula (1) is given in Figure 2. Function res(Z, a) has the following properties:

- it takes nonzero values if point Z lies inside the circle $\mathcal{C}$ and its image $Z \circ \tilde{h}(a)$ lies outside this circle;
- it tends to 0 if the point Z approaches the boundary of circle $\mathcal{C}$, and to 1 if the image $Z \circ \tilde{h}(a)$ approaches this boundary.

Observe that, when the length of the longest trajectory included in $\mathcal{C}$ is incremented as a result of moving the starting point Q towards the attractor, the largest residual term changes its value from 1 to 0. Consequently, $E_2(Q)$ is a continuous function of the point Q in the domain $X \backslash \mathcal{A}$. For a formal proof of this property see [8].

The continuity of the escape function can also be maintained by other residual terms, for example

$$\text{res}'(Z, a) = \begin{cases} \dfrac{R - \|Z\|}{\|Z \circ \tilde{h}(a)\| - \|Z\|} & \text{if } Z \in \mathcal{C} \text{ and } Z \circ \tilde{h}(a) \notin C, \\ 0 & \text{otherwise.} \end{cases}$$

In order to explain the advantages of formula (1), let us consider an IFS consisting of a single complex function $F(z) = z/c$. By definition, $F(z)$ is a contraction,

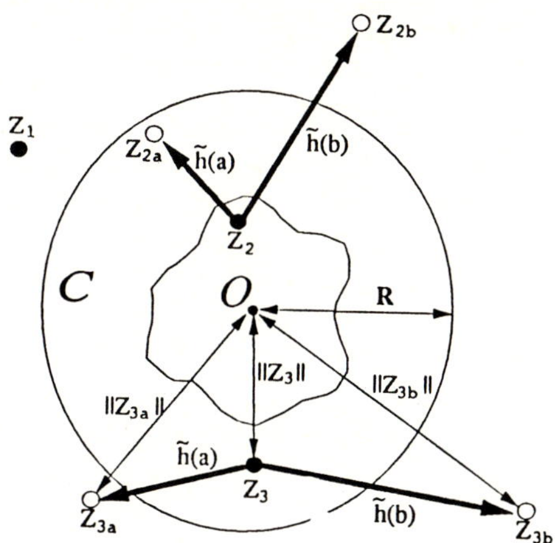

- $res(Z_1, a) = \text{res}(Z_1, b) = 0$, since $Z_1 \notin \mathcal{C}$,
- $\text{res}(Z_2, a) = 0$, since $Z_2 \circ \tilde{h}(a) = Z_{2a} \in \mathcal{C}$,
- $\text{res}(Z_3, a) > res(Z_3, b)$, since $\|Z_{3a}\| < \|Z_{3b}\|$.

Figure 2: Illustration of the residual terms $\text{res}(Z, a)$

thus $|c| > 1$. Given a circle $\mathcal{C}$ with radius R, and with center O in the origin of the coordinate system, the integer-valued escape-time function $E_1(z)$ is equal to:

$$E_1(z) = \max\{n \in \mathcal{N} : \|zc^n\| \leq R\}.$$

The symbol $\mathcal{N}$ represents the set of natural numbers (including zero), and the module of a complex number is identified with its norm, $|zc^n| = \|zc^n\|$. A continuous (and infinitely differentiable) extension of function $E_1(z)$ is:

$$E_2(z) = \max\{u \in \mathcal{R}^+ : \|zc^u\| \leq R\},$$

where $\mathcal{R}^+$ is the set of nonnegative real numbers. Obviously, the value $E_2(z)$ satisfies the equation:

$$\|zc^{E_2(z)}\| = R.$$

Consider point $Z = zC^{E_1(z)}$. By representing $E_2(z)$ as a sum $E_1(z) + \text{res}(Z)$, we obtain[1]:

$$\|zc^{E_1(z)+\text{res}(Z)}\| = \|Zc^{\text{res}(Z)}\| = R.$$

Note that $c = Zc/Z = F^{-1}(Z)/Z$, and take logarithms of both sides of the previous equation:

$$\log\|Z\| + \text{res}(Z)(\log\|F^{-1}(Z)\| - \log\|Z\|) = \log R.$$

[1]There is no need for specifying the second argument of function res, as the IFS under consideration consists of a single transformation.

Consequently,

$$\mathrm{res}(Z) = \frac{\log R - \log \|Z\|}{\log \|F^{-1}(Z)\| - \log \|Z\|}.$$

Although the above reasoning applies to a particular IFS, it justifies the use of function (1) also in other cases. In general, the distance between the origin of circle $\mathcal{C}$ and consecutive points in an escape trajectory tends to grow exponentially for large distance values. Consequently, function 1 minimizes first-order discontinuities in the escape-time function, yielding visually pleasing graphical representations.

The escape-time functions $E_1(Q)$ and $E_2(Q)$ are not defined inside the attractor $\mathcal{A}$, as one can find there an infinite sequence of points remaining in $\mathcal{A}$ and therefore remaining in the circle $\mathcal{C}$. In order to make the definition of the escape time computationally effective, we evaluate the escape trajectories up to a predefined maximum length M. The escape-time functions, limited in this way, can be computed in the entire space X using the following formulae:

$$\underline{E}_1(Q,M) = \begin{cases} 0 \ \text{ if } Q \notin \mathcal{C} \ \text{ or } \ M = 0 \\ 1 + \max_{a \in V}\{\underline{E}_1(Q \circ \tilde{h}(a), M-1)\} \ \text{ otherwise.} \end{cases}$$

$$\underline{E}_2(Q,M) = \begin{cases} 0 \ \text{ if } Q \notin \mathcal{C} \ \text{ or } \ M = 0, \\ \max_{a \in V}\{res(Q,a)\} & \text{if } Q \in \mathcal{C},\ M > 0 \text{ and } Q \circ \tilde{h}(a) \notin \mathcal{C} \text{ for all } a \in V, \\ 1 + \max_{a \in V}\{\underline{E}_2(Q \circ \tilde{h}(a), M-1)\} & \text{otherwise.} \end{cases}$$

It is intuitively clear that $\underline{E}_1(Q,M) = E_1$ for all points Q with the escape time $E_1(Q)$ less than M, since the recursive formula evaluates step-by-step the same trajectories as its non-recursive counterpart. Similarly, $\underline{E}_2(Q,M) = E_2(Q)$ for all points Q such that $E_2(Q) < M$. Rigorous proofs of these equalities can be carried out by induction on M.

Figure 3 visualizes the dragon curve from Figure 1 using the continuous escape-time function $\underline{E}_2(Q,M)$. The inverse functions are:

$$\begin{aligned} F_1^{-1} &= r(-45) \circ s(\sqrt{2}), \\ F_2^{-1} &= t(0,-1) \circ r(-135) \circ s(\sqrt{2}). \end{aligned}$$

The circle $\mathcal{C}$ has radius R equal to 5, and the limit M is equal to 20. The values of function $\underline{E}_2(Q,M)$ are interpreted as a height field.

5 Language-restricted IF's

A *language-restricted* iterated function system (LRIFS) is a quintuplet $\mathcal{I}_L = \langle X, \mathcal{F}, V, h, L\rangle$, where $X, \mathcal{F}, V$ and h form an "ordinary" IFS, and $L \subset V^\star$ is a language over the alphabet V.

Figure 3: The dragon curve visualized using the escape-time method

Consider a starting point P that belongs to the attractor $\mathcal{A}$ of the IFS $\mathcal{I}$, and let $\mathcal{A}_L(P)$ denote the image of P with respect to the transformations $h(L)$. The following inclusion holds:

$$\mathcal{A}_L(P) = P \circ h(L) \subset P \circ h(V^*) = \mathcal{A}.$$

Thus, the set $\mathcal{A}_L(P)$ generated by the LRIFS $\mathcal{I}_L$ with the starting point $P \in \mathcal{A}$ is a subset of the attractor $\mathcal{A}$. For example, consider an LRIFS $\mathcal{F}_L = \langle X, \mathcal{F}, V, h, L \rangle$, where:

- the space X is the plane,
- the IFS $\mathcal{F}$ consists of four transformations:

$$\begin{aligned} F_1 &= s(0.5) \\ F_2 &= s(0.5) \circ t(0, 0.5) \\ F_3 &= s(0.5) \circ r(45) \circ t(0, 1) \\ F_4 &= s(0.5) \circ r(-45) \circ t(0, 1), \end{aligned}$$

- the alphabet V consists of four letters a, b, c, d,
- the homomorphism h is defined by:

$$h(a) = F_1, \quad h(b) = F_2, \quad h(c) = F_3, \quad h(d) = F_4,$$

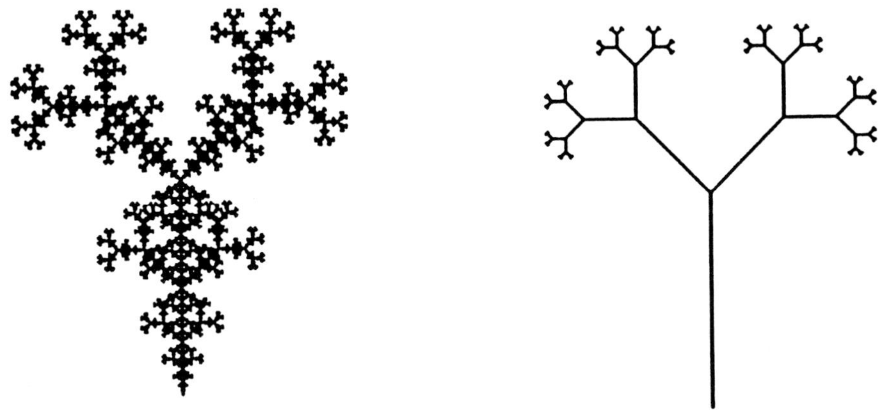

Figure 4: Attractor $\mathcal{A}$ and its subset $\mathcal{A}_L(P)$

- the language L consists of words in which no letter c or d is followed by an a or b. [2]

Figure 4 compares the attractor $\mathcal{A}$ of the IFS $\mathcal{I}$ with the set $\mathcal{A}_L(P)$ generated by the LRIFS $\mathcal{I}_L$ using the starting point $P = (0,0)$. Clearly, the branching structure of Figure (b) is a subset of the original attractor (a).

In general, the set $\mathcal{A}_L(P)$ depends on the choice of the starting point P. Nevertheless, if the language L is prefix extensible, $vL \subset L$, the smallest set $\mathcal{A}_L$ exists and can be found as $P_0 \circ h(L)$, where P_0 is the invariant point of the transformation $h(v)$. This results from the following inclusions, satisfied for any $P \in X$:

$$P_0 \circ h(L) = (\lim_{n\to\infty} P \circ h(v^n)) \circ h(L) = \lim_{n\to\infty} P \circ h(v^n L) \subset P \circ h(L).$$

The limits are calculated in the space of all closed nonempty bounded subsets of the space X with the Hausdorff metric [13].

By analogy with the "ordinary" IFS's, we call $\mathcal{A}_L$ the attractor of the LRIFS $\mathcal{I}_L$.

6 The Escape-time Method for LRIF's

While extending the escape-time method to LRIFS's, we will consider mirror images of words and use the following lemma.

Lemma. Consider an IFS $\mathcal{I} = \langle X, \mathcal{F}, V, h\rangle$, where all functions $F = h(a) \in \mathcal{F}$ are invertible. For any word $w \in V^\star$, the equality $\tilde{h}(w) = (h(w^R))^{-1}$ holds.

Proof. The set $\mathcal{F}$ forms a group of transformations with the operations of function composition and inversion, thus $(F_i \circ F_j)^{-1} = F_j^{-1} \circ F_i^{-1}$ for any $F_i, F_j \in$

[2]Thus, L is defined by the regular expression $L = (a \cup b)^*(c \cup d)^*$.

$\mathcal{F}$. Consequently, the following equalities are true for any word $w = a_1a_2\ldots a_n \in V^\star$:

$$\begin{aligned}\tilde{h}(w) &= \tilde{h}(a_1a_2\ldots a_n) = \tilde{h}(a_1)\circ\tilde{h}(a_2)\circ\ldots\circ\tilde{h}(a_n) = \\ &= (h(a_1))^{-1}\circ(h(a_2))^{-1}\circ\ldots\circ(h(a_n))^{-1} = \\ &= (h(a_n)\circ\ldots\circ h(a_2)\circ h(a_1))^{-1} = \\ &= (h(a_n\ldots a_2a_1))^{-1} = (h(w^R))^{-1}. \ \square\end{aligned}$$

The escape-time method for LRIFS's is based on the following extension of Theorem 1 from Section 4:

Theorem 2. Consider an LRIFS $\mathcal{I}_L = \langle X, \mathcal{F}, V, h, L\rangle$, and assume that the language L is prefix extensible, $vL \subset L$. Denote by $\mathcal{A}$ the attractor of the IFS $\langle X, \mathcal{F}, V, h\rangle$, and by $\mathcal{A}_L$ the attractor of $\mathcal{I}_L$. (a) If a starting point $Q \in X$ belongs to the attractor $\mathcal{A}_L$, then for any $n \geq 0$ there exists a word w in the prefix closure $\mathcal{P}(L^R)$ such that length$(w) \geq n$ and $\mathrm{Tr}(Q, w) \subset \mathcal{A}$. (b) If the point Q does not belong to $\mathcal{A}_L$, all trajectories $\mathrm{Tr}(Q, w)$ with $w \in \mathcal{P}(L^R)$ diverge to infinity as length$(w) \to \infty$.

Proof. (a) Let P_0 denote the invariant point of transformation $h(v)$. According to the definition of the attractor $\mathcal{A}_L$, there exists a word $y \in L$ such that $P_0 \circ h(y) = Q$. Since $P_0 = P_0 \circ h(v)$, the equality $P_0 \circ h(v^iy) = Q$ holds for any $i \geq 0$. Let i satisfy the inequality length$(v^iy) \geq n$, and $w = (v^iy)^R$. This word belongs to L^R and henceforth to $\mathcal{P}(L^R)$, has length greater than or equal to n, and maps point Q to the point $P_0 \in \mathcal{A}_L$:

$$Q \circ \tilde{h}(w) = Q \circ (h(v^iy))^{-1} = P_0 \in \mathcal{A}_L.$$

In order to show that the entire trajectory $\mathrm{Tr}(Q, w)$ is included in the attractor $\mathcal{A}$, let us consider an arbitrary partition of the word w into a prefix x_1 and a suffix x_2; thus $x_1x_2 = w$. From the equality

$$Q \circ \tilde{h}(w) = Q \circ \tilde{h}(x_1) \circ \tilde{h}(x_2) = P_0$$

it follows that

$$Q \circ \tilde{h}(x_1) = P_0 \circ (\tilde{h}(x_2))^{-1} = P_0 \circ h(x_2^R) \in P_0 \circ h(V^\star) = \mathcal{A}.$$

Since this argument holds for any $x \prec w$, we obtain:

$$\mathrm{Tr}(Q, w) = \{Q \circ \tilde{h}(x) : x \prec w\} \subset \mathcal{A}.$$

(b) Let $\mathcal{C}$ be an arbitrary circle enclosing the attractor $\mathcal{A}$, and R denote the radius of $\mathcal{C}$. We have to prove that if $Q \notin \mathcal{A}_L$, there exists a number $n \geq 0$ such that for any word $w \in \mathcal{P}(L^R)$ of length greater then or equal to n, the escape trajectory $\mathrm{Tr}(Q, w)$ is not entirely included in $\mathcal{C}$. Let D denote the distance between point Q and the attractor $\mathcal{A}_L$, and r_{max} be the largest Lipschitz constant found among the transformations $F \in \mathcal{F}$. Since $D > 0$ and $r_{max} < 1$, there exists a number

$n \geq 0$ such that $2Rr^n_{max} < D$. Consider an arbitrary word $w \in \mathcal{P}(L^R)$ with length$(w) \geq n$, and let $wy \in L^R$. Since $y^R w^R \in L$, there exist points $P_0, P \in \mathcal{A}_L$ such that $P_0 \circ h(y^R w^R) = P$. We decompose the last equality by introducing an intermediate point P':

$$P_0 \circ h(y^R) = P' \text{ and } P' \circ h(w^R) = P.$$

It follows that

$$P \circ \tilde{h}(w) = P' = P_0 \circ h(y^R) \in P_0 \circ h(V^\star) = \mathcal{A}.$$

The distance between points P and Q is at least D, and the Lipschitz constant of the composite transformation $\tilde{h}(w) = (h(w^R))^{-1}$ is at least r^{-n}_{max}, thus

$$d(Q \circ \tilde{h}(w), P \circ \tilde{h}(w)) \geq d(Q, P) r^{-n}_{max} \geq D r^{-n}_{max} > 2R.$$

Since $P \circ \tilde{h}(w) = P' \in \mathcal{A} \subset \mathcal{C}$, and the distance of $Q \circ \tilde{h}(w)$ from P' is greater than the diameter of $\mathcal{C}$, the point $Q \circ \tilde{h}(w)$ must lie outside of $\mathcal{C}$, or

$$\mathrm{Tr}(Q, w) \not\subset \mathcal{C}. \quad \square$$

Theorem 2 reveals an analogy between the escape trajectories of an LRIFS and an ordinary IFS. In both cases we find infinitely long trajectories confined to $\mathcal{A}$ if the starting point Q belongs to the attractor — respectively $\mathcal{A}_L$ or $\mathcal{A}$. For a point Q outside an attractor, all trajectories diverge to infinity as their length increases. However, in the case of an ordinary IFS we consider escape trajectories with respect to all possible words $w \in V^\star$, while in the case of an LRIFS the words w are confined to the prefix closure $K = \mathcal{P}(L^R)$.

As a result of this observation, we can extend the escape-time formulae from Section 4 to LRIFS's as follows:

$$\begin{aligned} E_{L1}(Q) &= \max_{w \in K}\{\mathrm{length}(w) : \mathrm{Tr}(Q, w) \subset \mathcal{C}\}, \\ E_L 2(Q) &= \max_{wa \in K}\{\mathrm{length}(w) + \mathrm{res}(Q \circ \tilde{h}(w), a) : \mathrm{Tr}(Q, w) \subset \mathcal{C}\}. \end{aligned}$$

In the recursive counterparts of these functions, the key issue is the selection of mappings $\tilde{h}(a)$ that can be used in each step. They correspond to the initial letters of the words in L^R, or to the single-letter words in K. We use the derivatives of the language K to find the initial letters in the next level of recursion. As previously, M limits the recursion depth.

$$\underline{E}_{L1}(Q, K, M) = \begin{cases} 0 \ \text{ if } Q \notin \mathcal{C} \text{ or } M = 0 \\ 1 + \max_{a \in K}\{\underline{E}_{L1}(Q \circ \tilde{h}(a), K//a, M - 1)\} \ \text{ otherwise.} \end{cases}$$

$$\underline{E}_{L2}(Q, K, M) = \begin{cases} 0 \ \text{ if } Q \notin \mathcal{C} \text{ or } M = 0, \\ \max_{a \in K}\{res(Q, a)\} & \begin{array}{l}\text{if } Q \in \mathcal{C},\ M > 0 \text{ and} \\ Q \circ \tilde{h}(a) \notin \mathcal{C} \text{ for all } a \in K,\end{array} \\ 1 + \max_{a \in K}\{\underline{E}_{L2}(Q \circ \tilde{h}(a), K//a, M - 1)\} \ \text{ otherwise.} \end{cases}$$

These formulae can be used for any language $K = \mathcal{P}(L^R)$, provided that L has the prefix property, as assumed in Theorem 2. The required operations on languages are particularly simple if L is regular. It can be then specified using a finite-state automaton, which reduces operations on infinite languages to the operations on their finite representations. Details are given in the following section.

7 The Application of Finite Automata

We start by recalling the necessary notions of the theory of finite automata. For the original presentation see [16].

A *nondeterministic finite-state (Rabin-Scott) automaton* is a quintuplet:

$$\mathcal{M} = < V, S, s_0, T, I >,$$

where:

- V is an alphabet,
- S is a finite set of states,
- $s_0 \in S$ is a distinguished element of S, called the initial state,
- $T \subset S$ is a distinguished subset of S, called the set of final states,
- $I \subset V \times S \times S$ is a state transition relation.

Instead of $(a, s_i, s_k) \in I$ we may write $(a, s_i) \rightarrow s_k$.

Finite state automata are commonly represented as directed graphs, with the nodes corresponding to states and arcs representing transitions. The initial state is pointed to by a short arrow. The final states are distinguished by double circles.

A word $w = a_1 a_2 \ldots a_n \in V^\star$ is *accepted* by the automaton $\mathcal{M}$ if there exists a sequence of states $s_0, s_1, s_2, \ldots, s_{n-1} \in S$ and $s_n \in T$ such that

$$(a_1, s_0) \rightarrow s_1, \quad (a_2, s_1) \rightarrow s_2 \quad \cdots \quad (a_n, s_{n-1}) \rightarrow s_n$$

Thus, w is accepted by $\mathcal{M}$ if there exists a directed path in the graph of $\mathcal{M}$ starting in the initial state s_0, ending in some final state s_n, and labeled with the consecutive letters of w.

The set of all words accepted by an automaton $\mathcal{M}$ is called the *language accepted by* $\mathcal{M}$, and denoted by $L(\mathcal{M})$.

It is known that the mirror image of the language $L(\mathcal{M})$ is accepted by the automaton

$$\mathcal{M}^{\mathcal{R}} = \langle V, S \cup \{s'_0\}, s'_0, \{s_0\}, I^R \rangle,$$

where:

$$I^R = \{(\epsilon, s'_0, s_k) : s_k \in T\} \cup \{(a, s_j, s_i) : (a, s_i, s_j) \in I\}.$$

Thus, the automaton $\mathcal{M}^{\mathcal{R}}$ can be obtained from $\mathcal{M}$ by:

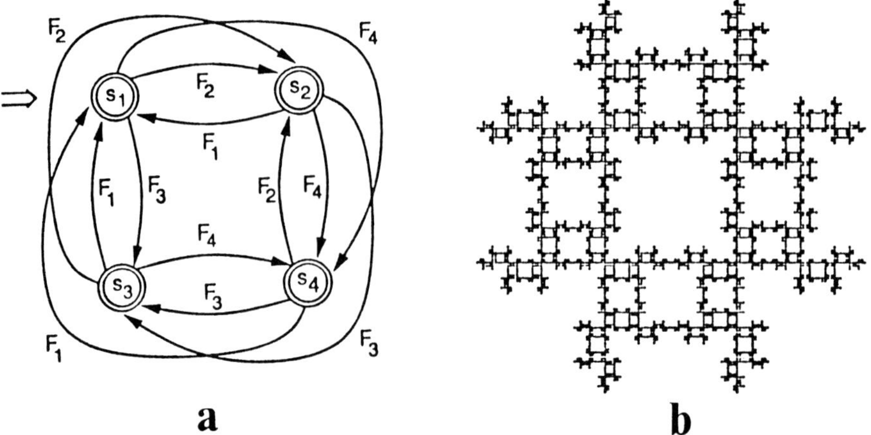

Figure 7: (a) The automaton $\mathcal{M}_2$ defining the language L_2, and (b) the attractor of the LRIFS $\mathcal{I}_2$

The language L_1 is defined using the finite automaton $\mathcal{M}_1$ shown in Figure 5a, and the corresponding attractor is given in Figure 5b. The automaton defining the language $K_1 = \mathcal{P}(L_1^R)$ is shown in Figure 6, with the transitions labeled using the inverse transformations of $\mathcal{F}_1$. Plate 1 visualizes the escape time function computed with a recursion depth limit $M = 20$, using a bounding circle $\mathcal{C}$ with radius $R = 5$. The escape time values are interpreted as indices to a color map, arbitrarily divided into several ramps. Plate 2 presents the same function as a height field.

Example 2. The LRIFS $\mathcal{I}_2$ considered in this example was described by Vrscay [17]. It uses the same set of transformations $\mathcal{F}$ and the labeling function h as $\mathcal{I}_1$, but the language L_2 is different. The automaton $\mathcal{M}_2$ defining L_2 and the resulting attractor are shown in Figure 7. The escape time functions are presented in Plates 3 and 4.

Example 3. The LRIFS $\mathcal{I}_3$, taken from [13], describes a leaf-like structure with the alternating and opposite branches. The set of transformations is specified below:

$$\begin{aligned}
F_1 &= s(0.5) \circ t(-0.002, 0) \\
F_2 &= s(0.5) \circ t(0.002, 0) \\
F_3 &= s(0.5) \circ t(-0.002, 0.13) \\
F_4 &= s(0.5) \circ t(0.002, 0.13) \\
F_5 &= s(0.42) \circ r(45) \\
F_6 &= s(0.2) \circ r(90) \circ t(-0.05, 0.05) \\
F_7 &= s(0.2) \circ t(-0.05, 0.05) \\
F_8 &= t(0.3, -0.3) \circ s(0.74) \circ t(-0.3, 0.3) \\
F_9 &= s(0.37) \circ r(-45) \circ t(0, 0.14) \\
F_{10} &= s(0.172) \circ r(-90) \circ t(0.05, 0.19) \\
F_{11} &= s(0.172) \circ t(0.05, 0.19)
\end{aligned}$$

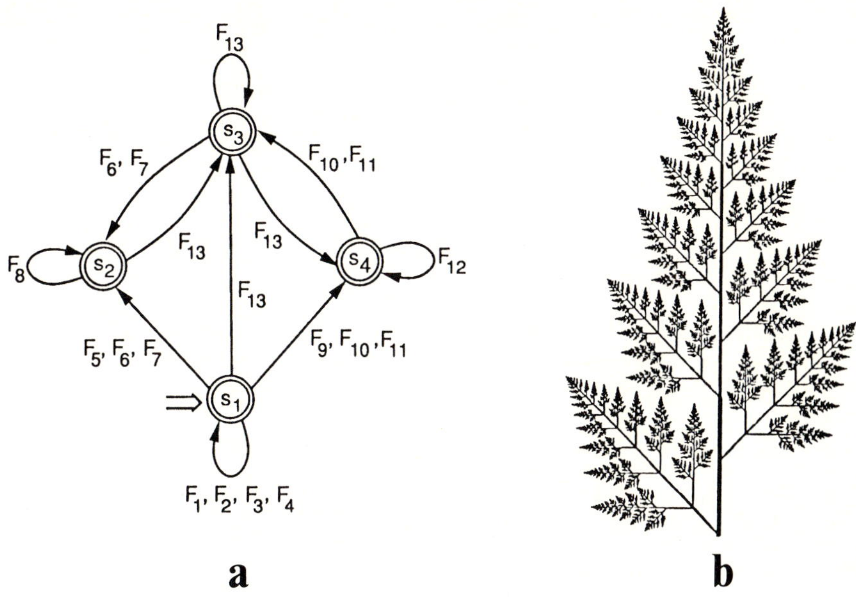

Figure 8: (a) The automaton $\mathcal{M}_3$ defining the language L_3, and (b) the attractor of the LRIFS $\mathcal{I}_3$

$$
\begin{aligned}
F_{12} &= t(-0.265, -0.405) \circ s(0.74) \circ t(0.265, 0.405) \\
F_{13} &= t(0, -1) \circ s(0.74) \circ t(0, 1)
\end{aligned}
$$

The automaton $\mathcal{M}_3$ defining L_3 and the corresponding attractor are shown in Figure 8. The automaton defining the language $K_3 = \mathcal{P}(L_3^R)$ is shown in Figure 9. The escape time functions are visualized in Plates 5 and 6. Plate 5 was generated using a continuous color ramp.

8 Conclusions

This paper presents methods for computing the escape-time functions of language-restricted iterated function systems. The LRIFS's generalize the ordinary IFS's by providing means for imposing restrictions on the sequences of transformations. The escape-time functions can be computed for any set of sequences described viewed as a prefix-extensible formal language L. The computation of the escape time involves the operations of finding the mirror image of L, determining the prefix language, and calculating the derivatives. They can be performed in a simple way if L is regular. All examples considered in this paper refer to this case. It is an open problem whether the use of non-regular languages can yield other attractors and visualizations.

One could raise a question, whether this paper applies computer graphics to visualize an important mathematical concept, or whether it merely employs

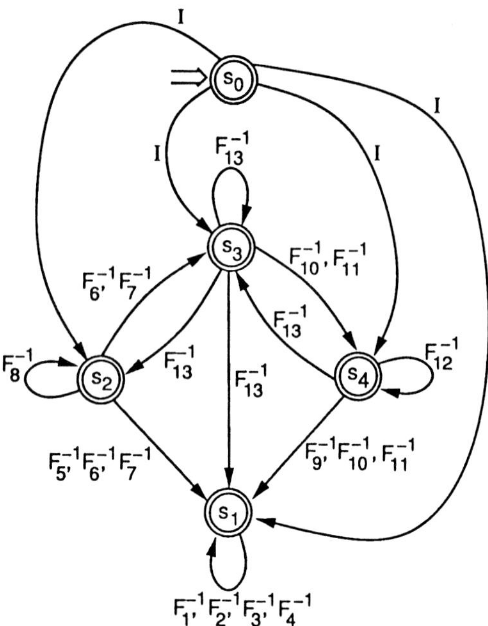

Figure 9: The automaton $\mathcal{M}_3^{\mathcal{RP}}$ defining the language $K_3 = \mathcal{P}(L_3^R)$

mathematics to create images for the sake of their visual appeal. Our motivation fell in both areas — we wanted to extend the mathematical concept of escape-time functions to LRIFS's, while realizing that it is primarily used for image synthesis. In addition, we found that the well-established theory of automata and formal languages had unexpected applications in computer graphics.

Acknowledgements

We would like to thank Dr. Dietmar Saupe for providing us with the program for rendering height fields. This research was sponsored by an operating grant and a graduate scholarship from the Natural Sciences and Engineering Research Council of Canada. The images were generated using facilities of the University of Calgary and the University of Regina.

References

[1] Ch. Bandt. Self-similar sets III. Constructions with sofic systems. *Monatsh. Math*, 108:89–102, 1989.

[2] M. F. Barnsley. *Fractals Everywhere*. Academic Press, 1988.

[3] M. F. Barnsley, J. H. Elton, and D. P. Hardin. Recurrent iterated function systems. *Constructive Approximation*, 5:3–31, 1989.

[4] M. A. Berger. Images generated by orbits of 2-D Markov chains. *Chance*, 2(2):18–28, 1989.

[5] J. Berstel and A. Nait Abdallah. Tétrarbres engendreés par des automates finis. Technical Report 89-7, Laboratoire Informatique Théorique et Programmation, Université P. et M. Curie, 1989.

[6] K. Culik II and S. Dube. Affine automata and related techniques for generation of complex images. Manuscript, University of South Carolina in Columbia.

[7] K. Culik II and S. Dube. Balancing order and chaos in image generation. Manuscript, University of South Carolina in Columbia.

[8] D. Hepting, P. Prusinkiewicz, and D. Saupe. Rendering methods for iterated function systems. In *Fractals in the Fundamental and Applied Sciences*, pages 183–224. Elsevier, 1991.

[9] J. E. Hutchinson. Fractals and self-similarity. *Indiana University Journal of Mathematics*, 30(5):713–747, 1981.

[10] B. B. Mandelbrot. *The Fractal Geometry of Nature*. W. H. Freeman, New York, 1982.

[11] H.-O. Peitgen and P. H. Richter, editors. *The Beauty of Fractals*. Springer-Verlag, Heidelberg, 1986.

[12] H.-O. Peitgen and D. Saupe, editors. *The Science of Fractal Images*. Springer-Verlag, New York, 1986.

[13] P. Prusinkiewicz and M. Hammel. Automata, languages, and iterated function systems. In J. C. Hart and F. K. Musgrave, editors, *Fractal Modeling in 3D Computer Graphics and Imagery*, pages 115–143. ACM SIGGRAPH, 1991. Course Notes C14.

[14] P. Prusinkiewicz and A. Lindenmayer. *The Algorithmic Beauty of Plants*. Springer-Verlag, New York, 1990. With J. Hanan, F. D. Fracchia, D. R. Fowler, M. J. M. de Boer, and L. Mercer.

[15] P. Prusinkiewicz and G. Sandness. Koch curves as attractors and repellers. *IEEE Computer Graphics and Applications*, 8(6):26–40, November 1988.

[16] M. O. Rabin and D. Scott. Finite automata and their decision problems. *IBM J. Res. Develop.*, 3:114–125, 1959.

[17] E. R. Vrscay. Iterated function systems: Theory, applications and the inverse problem. In *Proceedings of the NATO Advanced Study Institute on Fractal Geometry held in Montreal, July 1989*. Kluwer Academic Pulishers, 1990.

[18] T. E. Womack. Linear and Markov iterated function systems in fractal geometry. Master's thesis, Virginia Polytechnic Institute and State University, Blacksburg, Virginia, 1989.

Plate 1

Plate 2

Plate 3

Plate 4

Plate 5

Plate 6

1/f Noise and Fractals in Economic Time Series

R. F. Voss
IBM Thomas J. Watson Research Center
P. O. Box 218
Yorktown Heights, NY 10598 USA

Abstract

Many economic time series such as stock prices $\$(t)$ have spectral densities $S_{\$}(f)$ that vary as $1/f^2$ and increments $\Delta\$(t)$ with $S_{\Delta\$}(f) \propto constant$, indicating they closely follow the *efficient market* hypothesis and mimic a random walk or Brownian motion. For such quantities, a knowledge of the past is of no help in predicting the future. Measurements of $S_{|\Delta\$|}(f)$ for the *volatility* $|\Delta\$(t)|$, however, show $S_{|\Delta\$|}(f) \propto 1/f$ indicating long range fractal correlations that may be of some use in forecasting.

1 Introduction: Chaos, Fractals, and Economics

Many of Mandelbrot's fractal concepts were first introduced in the field of economics [BBM 63-71] before the name *fractal* was invented in 1975 [BBM 75-82]. Although Mandelbrot established the importance of Lévy stable distributions (not just Gaussian) and the *Noah* (large excursions from average) and *Joseph effects* (long time correlations) in economics more than 20 years ago, tt has only been in recent years that economists have shown great interest. Fractal geometry and chaos theory are now widely perceived as offering some hope for characterizing and understanding the seemingly random behavior of economic variables [PM 90-91]. Many attempts have been made to both model economic systems with chaotic non-linear dynamics of a few degrees of freedom and to estimate strange attractor dimensions from economic time series [Econ 88-91]. It has recently been shown, however, that the standard technique of estimating an embedding dimension, can not distinguish between the chaotic strange attractor of a low dimensional dynamic system (with few degrees of freedom) and a high dimensional random process such as fractional Brownian motion [OP 89],[JT 91]. Although Mirowski gives a pessimistic review of the direct relevance of chaos theory to economic forecasting, he is adamant about the relevance of Mandelbrot's fractal concepts to economics [PM 90-91].

The measurements reported here support this view. Standard *auto-correlation* and *spectral density* techniques from the physics and mathematics of random processes are applied in a new manner to price records. They demonstrate for the first time the existence of $1/f$ noise and corresponding fractal correlations over more than 3 decades of time scales in economic variables.

2 Random Processes, Spectral Density, and Autocorrelation

A number of standard techniques for characterizing random processes and their correlations are familiar to physicists and mathematicians [Fluct]. The *autocorrelation* or *pair-correlation* function $G(\tau)$ is a quantitative measure of how the fluctuations in a quantity $X(t)$ are correlated between times t and $t+\tau$:

$$G(\tau)=\langle X(t)\,X(t+\tau)\,\rangle$$

where the brackets $\langle\ldots\rangle$ denote sample or ensemble averages.

The spectral density $S(f)$ also provides information about the time correlations of $X(t)$. If $X(f)$ is the Fourier coefficient of $X(t)$ at frequency f,

$$X(f)\propto\int X(t)\,e^{-2\pi i f t}dt$$

then

$$S(f)=|X(f)\,|^2/\Delta f$$

where Δf is the effective bandwidth of the Fourier integral.

$S(f)$ and $G(\tau)$ are not independent. In most cases they are related by the Wiener-Khintchine relations [Fluct]:

$$S(f)\propto\int G(\tau)\cos 2\pi\tau\,d\tau$$

and

$$G(\tau)\propto\int S(f)\cos 2\pi f\,df.$$

Fast Fourier Transform (FFT) algorithms allow efficient computation of $S(f)$ directly from sample sequences and the estimation of $G(\tau)$ from $S(f)$ via the 2nd Wiener-Khintchine relation above.

Figure 1 displays samples of typical random functions and their spectral densities. The *white noise* $w(t)$ in Fig. 1(c) is the most random. It is characteristic of a process that has no correlations in time. The future is completely independent of the past and its $G(\tau)$ has the form $G(\tau)\propto\delta(\tau)$. Consequently, its spectral density $S(f)=constant$ with equal power at all frequencies f, like white light.

The integral of white noise $w(t)$ (or the summation of random increments) produces a *Brownian motion* $X(t)=\int w(t)\;dt$ or random walk as shown in Fig.1(a). Thus, $X(t)$ corresponds to the random diffusion of particle and the average distance traveled in a time T obeys the usual diffusion law

$$\Delta X(T)=\langle|X(t+T)-X(t)\,|^2\rangle^{1/2}\propto T^{1/2}.$$

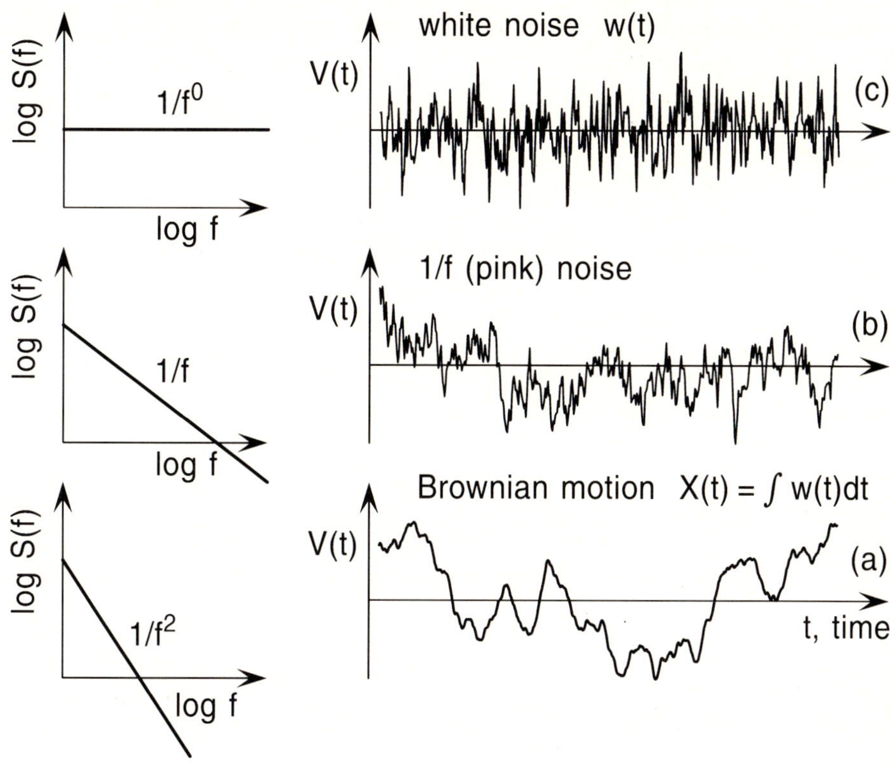

Figure 1: Typical noises and their spectral densities, $S(f)$.

Although the appearance is much more correlated and $S(f) \propto 1/f^2$, whether $X(t)$ increases or decreases in the future is independent of its entire past. Both the white noise $w(t)$ and its integral, Brownian motion, are examples of true random processes with no (or trivial) dependence on the past.

Many physical processes and their mathematical models are characterized by a single correlation time τ_0. In this case, $G(\tau) \approx constant$ for $\tau \ll \tau_0$ and $G(\tau) \approx 0$ for $\tau \gg \tau_0$. $S(f)$ then takes a typical Lorentzian form that varies between white noise $S(f) \propto constant$ for $f \ll 1/\tau_0$ and a random walk $S(f) \propto 1/f^2$ for $f \gg 1/\tau_0$.

3 Fractional Brownian Motion (fBM) and 1/f Noise

Although the *white noise* in Fig.1(a) and *Brownian motion* in Fig.1(c) are well understood mathematically and physically, they are characteristic of relatively few naturally occurring fluctuation phenomena. Many natural time series look much more like the $1/f$ *noise* or *pink noise* in Fig.1(b). A wide variety of measured quantities from electronic voltages and time standards to meteorological, biological, traffic and musical quantities show measured $S(f)$ varying as $1/f^\beta$ with $\beta \approx 1$ over many decades [RFV 75-88],[Fluct],[BBM 63-71]. Such quantities represent fractal or *scaling* processes in time [BBM 75-82]. In most cases, the physical reason for this behavior remains a mystery.

The most effective mathematical model of such behavior is *fractional Brownian Motion* or fBM as developed by Mandelbrot and Wallis [BBM 69]. fBM is a scale-independent extension to Brownian Motion that allows the description and modelling of processes with infinite-range dependence on their past. Most approximations to random fractals are based on fBM [BBM 82],[RFV 85-88]. It was called *fractional* as opposed to *fractal* since its origin preceded the name fractal by more than a decade. A fBM process $X_H(t)$ is specified by the single parameter H in the range $0 < H < 1$ such that

$$\Delta X_H(T) = \langle |X_H(t+T) - X_H(t)|^2 \rangle^{1/2} \propto T^{H/2}.$$

and a corresponding

$$S_H(f) \propto 1/f^{\beta} \quad \text{where} \quad \beta = 2H + 1.$$

Thus, as $H \to 0$ the fBm $X_H(t) \to 1/f$ noise. For $H=1/2$, $X_H(t)$ becomes normal Brownian motion with the diffusive $\Delta X \propto t^{1/2}$.

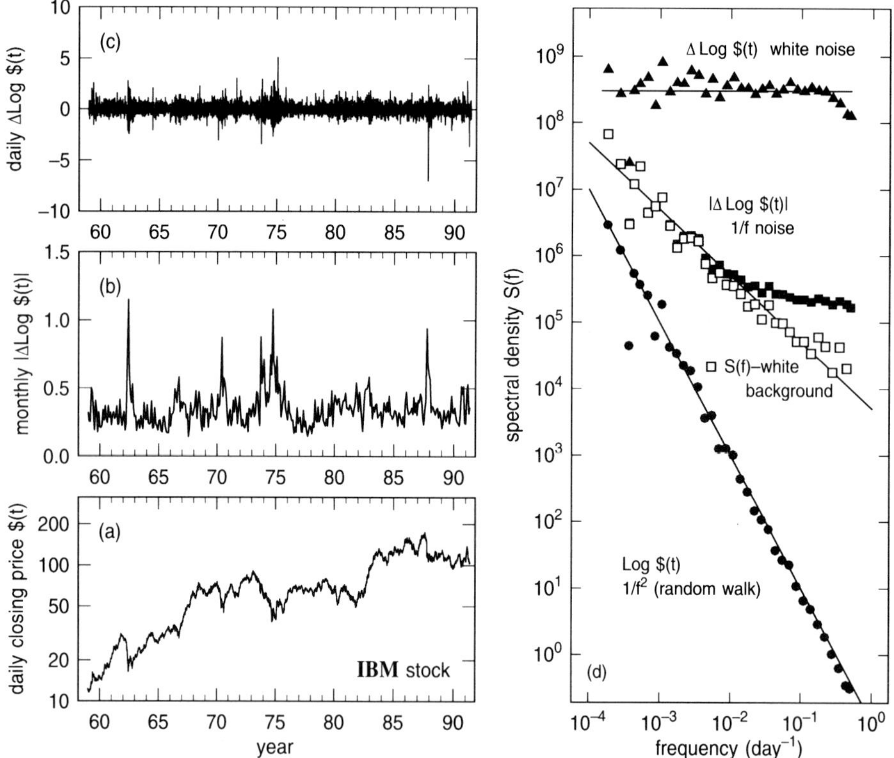

Figure 2: Daily IBM stock closing prices $\$(t)$, price changes $\Delta\$(t)$, and volatility $|\Delta\$(t)|$, and the corresponding measured spectral density $S(f)$.

4 Analysis of Prices, Price Changes, and Volatility

The above analysis methods can be easily applied to economic data. Figure 2 demonstrates their application to the daily closing price $\$(t)$ of IBM stock (adjusted for stock splits) from 1959 to early 1991. Fig. 2(a) shows the wide variations in $\log \$(t)$ over this time span. Here, the use of $\log \$(t)$ rather than $\$(t)$ effectively normalizes the daily variations relative to the current price. The corresponding measured $S_{\$}(f)$ varies almost exactly as the $1/f^2$ of a true random walk.

The daily price changes $\Delta\$(t)$ shown at the top in Fig. 2(c) have a *white* $S_{\Delta\$}(f)$. Both Fig. 2(a) and (c) show the expected behavior for the *efficient market* hypothesis where all investors have access to the same information at the same time. Such price changes are, therefore, unpredictable and independent of their past.

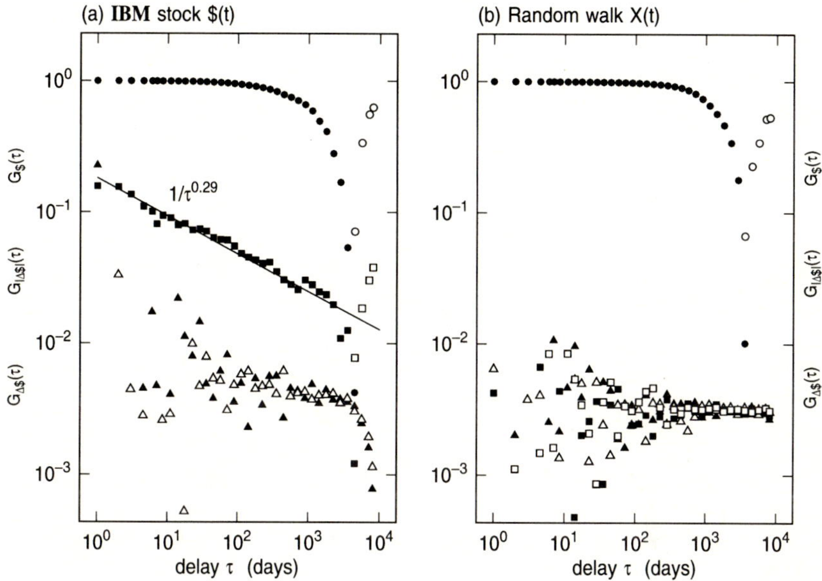

Figure 3: Autocorrelation analysis of the IBM stock time series from Fig. 2 compared with a pseudo-random walk. Positive $G(\tau)$ components are shown in solid symbols while negative components are shown with open symbols.

A careful comparison of the price changes in Fig. 2(c) with the white noise of Fig. 1(c), however, reveals a noticeable difference (as emphasized by Mandelbrot [BBM 75-82], visualization of both raw and processed data can play an important role in scientific understanding). The IBM stock price changes show a slowly varying envelope to the rapidly varying changes. These envelope variations are not found in a true random walk. Figure 2(b) shows the monthly average magnitude of the price changes $|\Delta\$(t)|$ or *volatility*. $|\Delta\$(t)|$ tracks the slow envelope changes in Fig. 2(c) and has a measured $S_{|\Delta\$|}(f) \propto 1/f$ at low f with a white noise limit at high f. Under the assumption that

$|\Delta\$(t)|$ is the sum of independent white noise and 1/f processes we may subtract the white noise limit from $S_{|\Delta\$|}(f)$ to show only the long-range correlations (open squares in Fig. 2(d)). Here the non-trivial fractal correlations extend over more than 3 decades.

Autocorrelation analysis $G(\tau)$ of the time series in Fig. 2 are shown in Fig. 3(a), while (b) shows the corresponding estimates for a random walk generated as the sum of pseudo-random increments. In both cases $G_{\$}(\tau)$ is essentially constant while $G_{\Delta\$}(\tau)$ is very small except where $\tau \approx 0$. A true random walk would have $G_{\Delta\$}(\tau)=\delta(\tau)$. Once again the differences appear in the volatility $|\Delta\$(t)|$ and the IBM stock $G_{|\Delta\$|}(\tau)$ shows a non-trivial scaling behavior (power-law) over more than 3 decades.

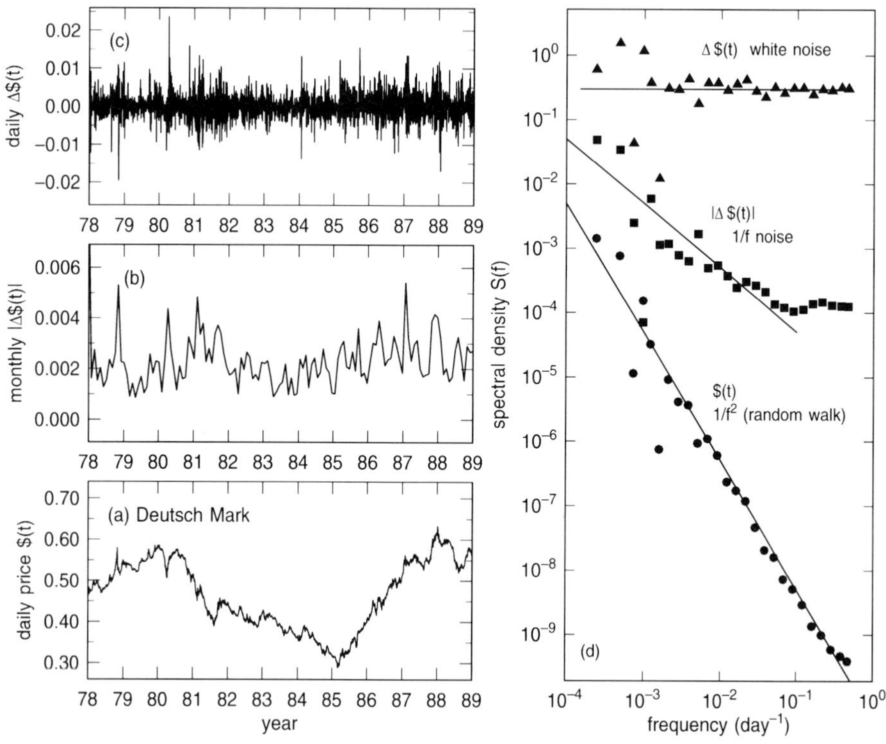

Figure 4: Daily currency prices of the Deutsch Mark *vs* the U.S. dollar (a) $\$(t)$, the price changes (c) $\Delta\$(t)$, and (b) volatility $|\Delta\$(t)|$ from 1978 through 1989.

A similar procedure may be applied to records of currency prices. One recent record of the Deutsch Mark vs the US dollar as shown in fig. 4 gives similar results. The direct prices have $S_{\$}(f) \propto 1/f^2$ characteristic of a random walk and $S_{\Delta\$}(f) \propto constant$ characteristic of a white noise. The volatility $|\Delta\$(t)|$ however, shows the characteristic $1/f$ dependence of long- range fractal correlations.

5 Conclusions

As demonstrated above, the common *spectral* and *autocorrelation* techniques can provide useful new information about economic time series.

A typical economic time series (such as daily IBM stock or Deutsch Mark currency prices) follows the *efficient market* hypothesis. The prices $\$(t)$ show $S_{\$}(f) \propto 1/f^2$ and the price changes $\Delta\$(t)$ show $S_{\Delta\$}(f) \propto constant$. These quantities mimic a random walk and its increments where a complete knowledge of the past is of no help in predicting the future.

The *volatility* $|\Delta\$(t)|$ on the other hand, shows $S_{|\Delta\$|}(f) \propto 1/f$ characteristic of long range fractal correlations and many naturally occurring $1/f$ *noises*. Here, correlations extend over all time scales and a knowledge of the past is helpful in predicting future volatility, but does not forecast the direction of the changes.

These measurements, however, do not specifically address the question of the relevance of chaos theory to economics [PR 90-91], [Econ 88-91]. Most natural $1/f$ noises are believed to be random processes, not chaos from a dynamical system of a few degrees of freedom [RFV 79]. In fact, it is extremely difficult to generate $S(f) \propto 1/f$ from non-linear iterated process of a few degrees of freedom.

References

[BBM 63-71] B.B. Mandelbrot,

New Methods in Statistical Economics, Journal of Political Economy, 421-440, October 1963.

... and J.W. van Ness, *Fractional Brownian motion, fractional noises and applications*, SIAM Review 10, 422-437, 1968.

Long-run Linearity, Locally Gaussian Process, H-spectra, and Infinite Variance, International Economic Review, 82-111, February 1969.

... and J.R. Wallis, *Some long-run properties of geophysical records*, Water Resources Research 5, 321-340, 1969.

When can price be arbitraged efficiently? A limit to the validity of the Random Walk and Martingale models, Review of Economics and Statistics, 225-236, August 1971.

[BBM 75-82] B.B. Mandelbrot,

les Objects Fractals, Flammarion, Paris, 1975.

Fractals: Form, Chance, and Dimension, W. H. Freeman and Co., New York, 1977;

The Fractal Geometry of Nature, W. H. Freeman and Co., New York, 1982.

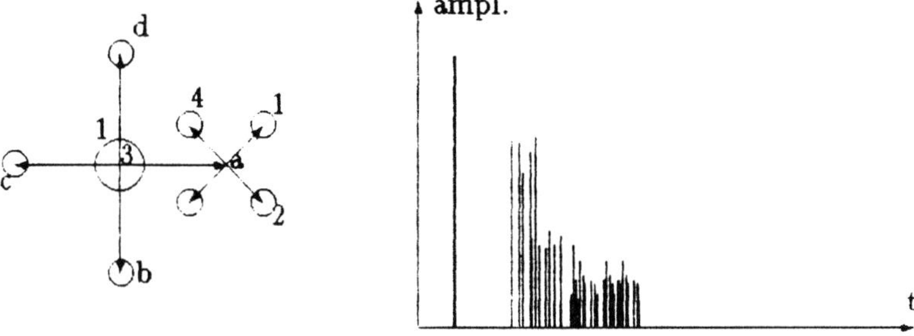

Figure 1: Waterdrop explodes in a pan with hot oil, the drops fall back into the oil and explode again until they are vaporised.

- Walking on the ice of a frozen lake.
- A drop of water in a pan with hot oil.
- The chaotic noise of an African market.
- Breaking material like metal or glass [PIE86].
- The noise of distant thunder.
- Waves breaking onto a beach have physically a self-similar structure.

These are just some examples and the reader is urged to find other good examples.

We pose the question and ask if an analogon between a fractal image and a fractal sound exists. Or, is it possible to create fascinating sound patterns as is done in computer graphics by using mathematical concepts like selfaffinity, selfsimilarity and iteration. Some people have already made experiments in the case of generating 'music' by mathematical methods like fractals [DOD86] cellular automatons [FUR91] or statistically [VOS75] but fractal music won't be the subject of this article.

2 Understanding Audio Fractals

2.1 Intuitiv Understanding

If we observe the behaviour of water in a pan with hot oil we see drops of water exploding and producing smaller drops which also explode. In figure 1.1 a model of this behaviour is drawn and the timeseries of the sound where every vertical line describes one single explosion. The length of this lines is equal to the loudness (amplitude) of the explosion or so called *event*.

The mathematical modellation of this system is quite easy. Thus we can write a small computer-programme to generate a sample with the specified behaviour which we can then listen to.

2.1.1 How to generate a Sample

If we want to make something audible by digital computers we should firstly study how we can bring the computer to play a sound.

A signal can be described as a timeseries

$$S = x(t) = \{x_0, x_1, \ldots, x_N\} \ . \tag{1}$$

This timeseries $S = x(t)$ will be translated by a digital-analog converter. The converter interpolates the discrete values and generates a voltage controlled signal which can be made audible by a normal HIFI-amplifier.

The systems described in this article generate a discrete series of values x_i equidistant in time:

$$x(t) = \{x_0, x_1, \ldots, x_N\} \;\; x(t) \in \{-2^{15} \ldots -1, 0, +1 \ldots 2^{15}\} \ .$$

The timeseries $x(t)$ is generally called a *sample*. These values can be made audible by converting them from a digital to an analogous Signal. This conversion is done by interpreting these values with a speed of 44100 values per second. This speed is called *Samplerate*. The maximum resolution is for some reasons [WAT 88] equal to a frequency of 22050 Hertz and thus covers the range of the human ear's perception(about $20 - 18000$Hz).

There is one special case of timeseries $x(t)$ which is interesting for our purpose:

$$x(t_0) = \int_{-\infty}^{\infty} x(t)\rho(t - t_o)dt \ .$$

This Integral is called *Dirac-Impulse*. If we watch the Fourier transformed function of it, we see that it contains all frequencys with the same amount of energy. The *Dirac-Impulse* has been succesfully used in the first system (Direct Generation in Time).

Further Reading on digital audio systems [WAT88].

2.1.2 Whats an Event ?

One explosion of water in oil could be described as the timeseries

$$E = \{\ldots 0, 0, e_i, e_{i+1}, \ldots, e_{m-1}, 0, 0, \ldots\} \ . \tag{2}$$

E will be called *event* and e the *eventfunction*. Now we can mix the events into a *sample* at different positions in time by simply adding the values of E at different positions in S (1). We are also able to scale E by any value in order to get any amplitude we need: $E_j = v \cdot E$. for $v < 1$ the event will be less loud than E.

3 Direct Generation in the Time Domain

Starting with a vector $\gamma_c = (t, l, v, \nu)$, we generate further vectors γ_i and collect them in the set

$$\Gamma = \{\Gamma_0, \Gamma_1, ..., \Gamma_k, ...\} \ . \tag{3}$$

The explicit system to generate Γ is:

$$\gamma_{i+1}^c \in \Gamma_0 \ . \tag{4}$$

$$\Gamma_{i+1} = \{\gamma_{i+1}^{j+wn} = f_j(z_w) | \forall z_w : z_w \in \Gamma_i\} \ . \tag{5}$$

$$f_j = (t + da_j, lb_j, vc_j, \nu d_j) \ . \tag{6}$$

a_j, b_j, c_j and d_j are constant.

t describes an abstract position in time where an event E should begin.

l is an abstract value, intuitivly meaning length.

v is more concrete and gives the amplitude for our event E.

ν is simply the motion of γ in time. Note that n could be also negative, which means a reverse order of time.

E is the event as descriped before (2).

Thus we firstly have to generate Γ_1 which contains n elements and then Γ_1 which already contains n^2 elements and n^k elements in the k-th generation.

Naturally this process is infinite, so we must restrict the generating process to get a finite set Γ. For this purpose we can use 2 possible limits. As you have already noticed in equation (3) we have marked Γ_k which will be our set after k generations. So we can say that processing will be stopped at the k-th generation.

But this is not the best way. As we see in equation (6), the next amplitude value vc_j for $c_j < 1$ could be used as a limit when v falls under the perceptual limit of the ear or under the limit of our discretisation step for the amplitudes $v < 1$.

After the set Γ has been generated we will now interpret this set and generate the timeseries S. The value t of each $\gamma_k^i \in \Gamma_i$ will be interpreted as the moment when an event E must be executed, $T(g)$ represents the function which extracts the value t of g and $V(g)$ extracts the amplitude value v.

$$X_j = \sum_{j=1}^{\|\Gamma_k\|} e(T(\gamma_k^i) - j) \cdot V(\gamma_k^i) \ .$$

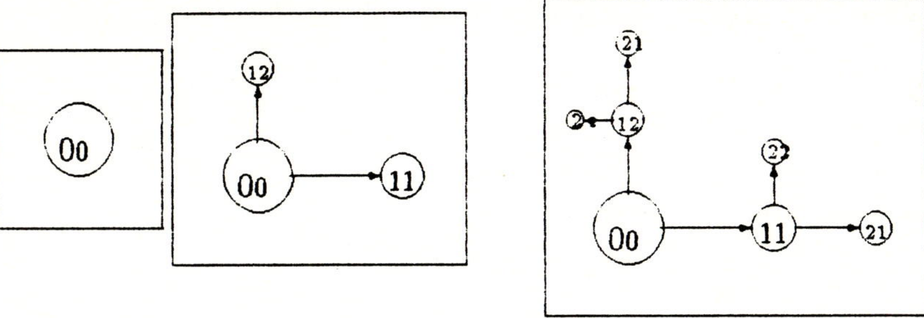

Figure 2: *from left to right:* Initial, first and second generation of a two branch system where the circle 0_0 generates in the first generation two circles 1_1 and 1_2. These two circles generate in the second generation the circles 2_1 and 2_2. The number G of G_D denotes the generation, the second number the daughter D .

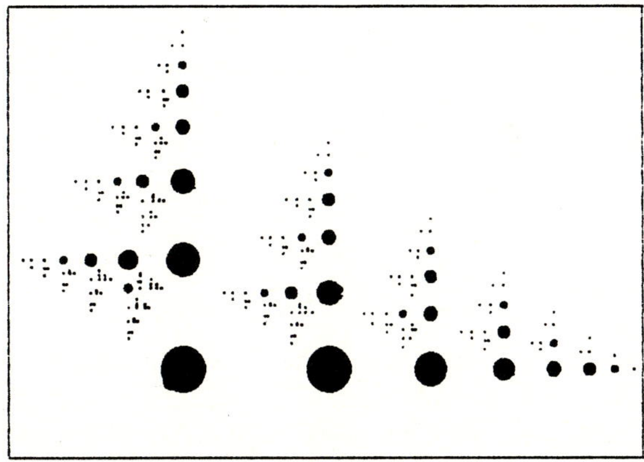

Figure 3: After 7 Generations we obtain this picture, following the generation rule of the previous figure (3.1).

3.1 Results

The sounds which can be obtained are very much dependant on the Eventfunction E. The experiments were made with $E = \{\ldots, 0, 0, 2^{15}, 0, 0, \ldots\}$ which is an approximation of the Dirac-impulse discussed earlier. The resulting sounds are similar to frying in oil, cold ice cubes in water and the noise of a longdistant AM radiotransmission. Other implementations of E will be tested in future. The problem of this model has been the non user friendly input of the constants which determine the sound. It's rather difficult to know beforehand the resulting sound.

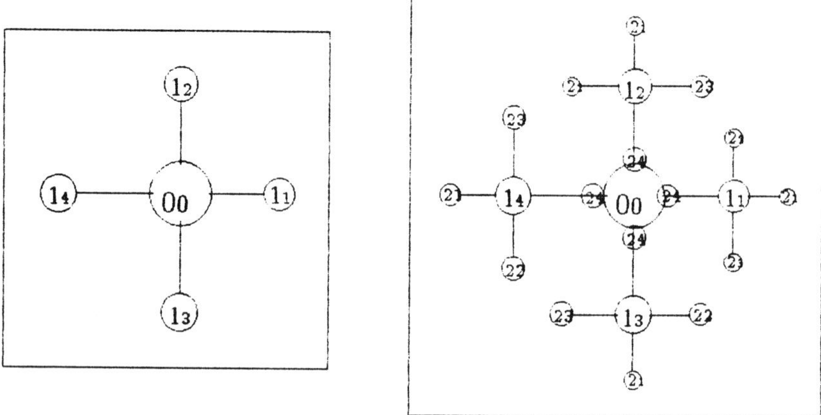

Figure 4: Another example but with 4 branches into all directions.

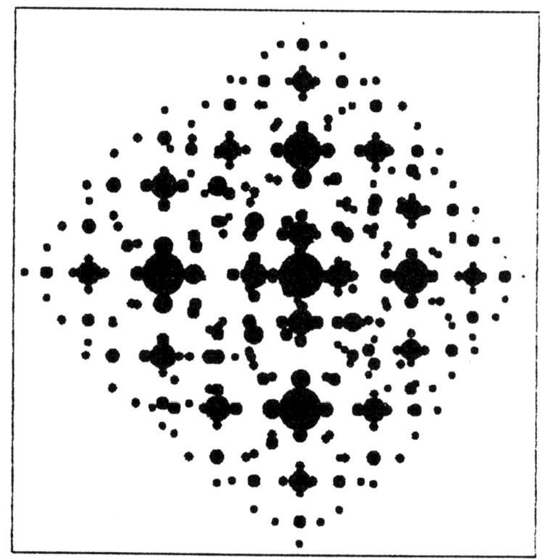

Figure 5: After 5 Generations we obtain this picture, following the generation rules of the previous figure (3.1).

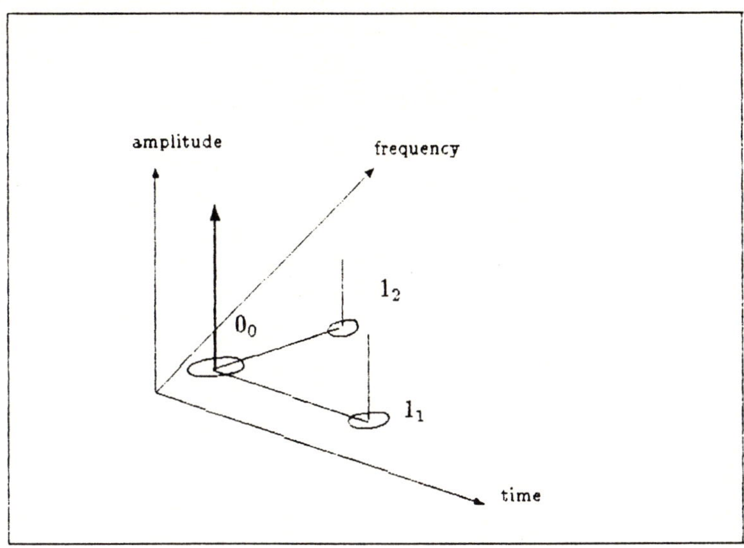

Figure 6: The interpretation of a mother circle and two daughters. The amplitude is determined by the radiant of every circle. The x-position gives the timeposition and the y-position the frequency (or pitch) of an Event. Low frequency is found near the time-axis.

4 Fractal Pictures into Fractal Sounds

4.1 Motivation

The first system isn't very easy to handle and you have to invest much time to obtain goodsounding results. I had to rethink the user-interface for this system which has been purly numerical so far. The best would have been a graphical interface, where the graphics correlate to the sound produced:

simple image $\rightsquigarrow$ simple sound
complex image $\rightsquigarrow$ complex sound
chaotic image $\rightsquigarrow$ chaotic sound
noisy image $\rightsquigarrow$ Noise

4.2 From Frequency Domain to Time Domain

In the following section I'll introduce a system which firstly generates an image built on circles which themselves generate further circles. This image will be stored in a large 2-D matrix M and then be transformed by the function Ξ:

$$\Xi(M) = S = \{x_1, x_2, \ldots, x_N\} \ . \qquad (7)$$

4.3 Generation of Circles

The generation of the picture is as follows(Fig. 3.1): A circle K represented by the vector (x, y, r, ϕ) generates n Circles K_i with varied parameters using the constants $(L, \alpha)_i = (L_i, \alpha_i)$.

$$K_i = (x + a_i \sin(\phi + \alpha_i), y + a_i \cos(\phi + \alpha_i), l \cdot L_i, \phi + a_i) \tag{8}$$

These constants are very easy to handle on a computersystem because they are made visible by lines with a specified length l and angle α) which can be transformed by interactive direct manipulation interfaces [SHN83].

The circles will be visualized on a computerscreen by drawing $(x, y, R \cdot r)$. R is a constant factor, which draws the generated circles with a fixed ratio to the internal radiant r. When generating further circles the limitations to obtain a finite set of circles are as these described in the first system. In our Matrix M we store the value r at position (x, y)

$$M = \begin{vmatrix} m_{11} & \dots & m_{I1} \\ \vdots & \ddots & \vdots \\ m_{1J} & \dots & m_{IJ} \end{vmatrix} \quad M = \begin{vmatrix} & \vdots & \\ \cdots & r_{xy} & \cdots \\ & \vdots & \end{vmatrix} \tag{9}$$

I, J are the limits of Matrix M. Circles which land outside of M $(x, y < 1 \wedge x > I, y > J)$ will simply be ignored.

4.4 Interpretation of Circles Matrix

The following equation gives the interpretation of this matrix to obtain a sample. The crucial element is the function $f(i)$ which determines the behaviour of the sound and it is left to the reader how $f(i)$ could be set so as to obtain a reverse Fourier transformation with I-frequencys.

$$\Xi(M) = \{x_j = \sum_{i=1}^{I} sin(\varphi_i(j) + f(i)) \cdot m(i, \mathrm{int}(\frac{j}{A}))\}, j \in \{1, 2, \dots, AJ\} \tag{10}$$

$\varphi_i(j)$ takes care that the overlapping timeslices[WAT88] are correctly in phase. The timeslices are overlapped to prevent hard entry of the sinewaves, because the function will no longer be steady at the moment of entry, thus generating unwelcome overtones. The frequency function $f(i)$ used is $f(i) = 33 \cdot \log_{12} \frac{i}{12}, i \in \{1 \dots 128\}$. The generating system now can be turned up by 1° and, after generation (8) and interpretation (10), we get a second timeseries, which is slightly different in frequency and time structure. In this way we can easily generate a stereosignal. The quality of the stereoeffect is not adequate for real room-acoustic, yet it is interesting in its psychological effect. You get the feeling of a strange structured imaginary room while the source of the sound seems to move around.

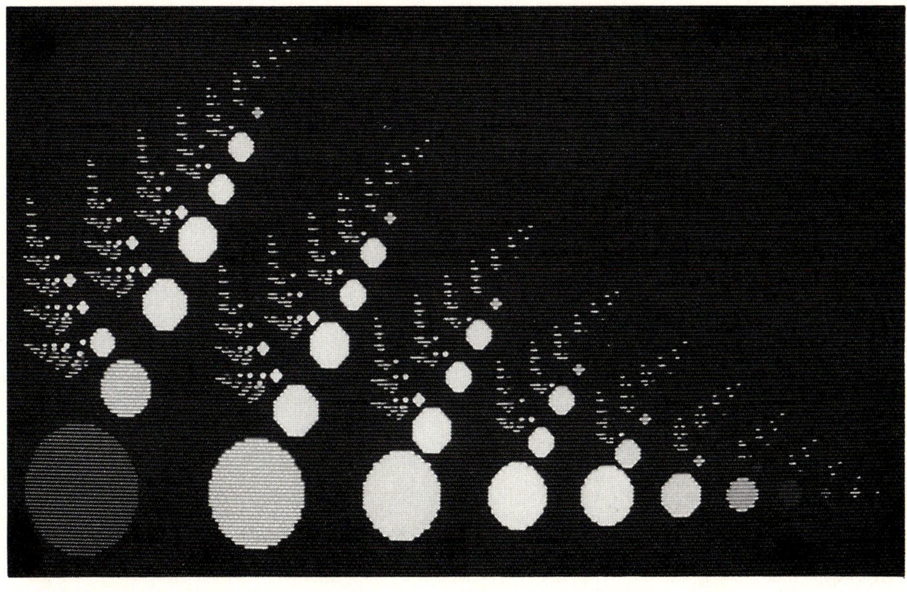

Figure 7: This pictures creates a sound similar to running water. The effect is very much dependant on the timescale and the eventfunction E (see formular 2). Using a long timescale like 30 seconds we obtain a strange music.

4.5 Result

This system is much easier to handle than the first system. It's quite easy to imagine the sound while manipulating the image in realtime. The system has been utilized in a big commercial application which handles the sounds of musical and film studio productions.

5 Examples

Let's have a look at some nice pictures which also generate nice sounds. (It's a pitty that this book can't play these sounds...)

References

[DOD86] Charles Dodge and Curtis R.Bahn, *Musical Fractals*, Byte, June 1986 pp 185-196

[FUR91] Furukawa, Kiyoshi, Personal playback performance of *'Splendent meridian II'*,Fractal Composition with Cellular Automaton, Musikhochschule Hamburg 1991

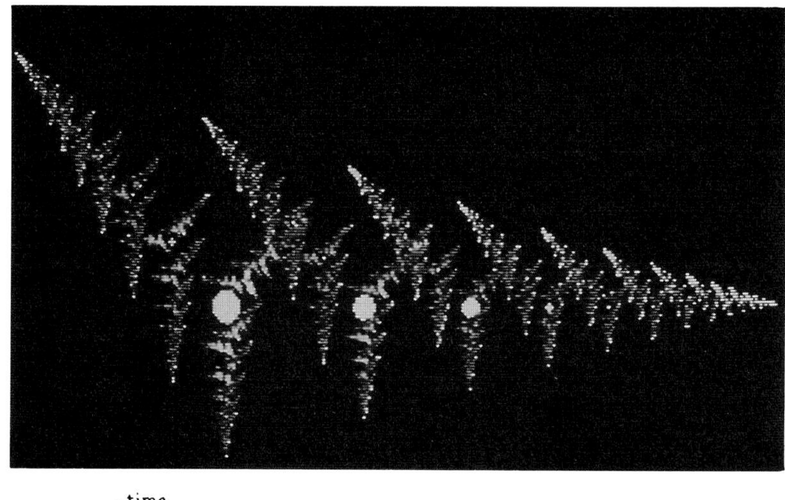

Figure 8: Here we could hear lots of action. Many small events could be distinguished.

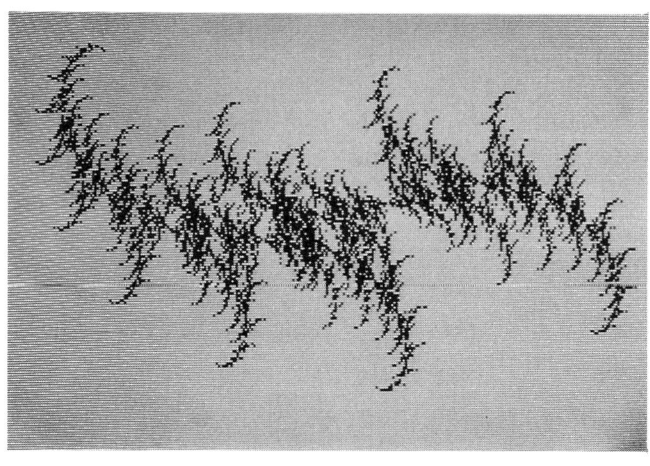

Figure 9: The events start before the main event, as with the sound of distant thunder.

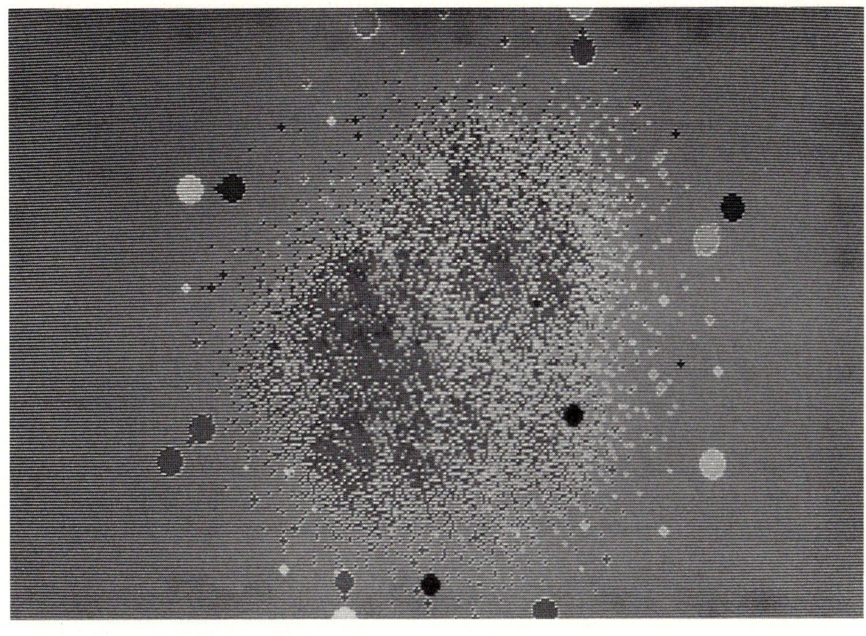

Figure 10: A chaotic set of circles generates a chaotic sound.

[MAN83] B.B.Mandelbrot: *The Fractal Geometry of Nature*,W.H. Freeman and company, New York1983

[PEI88] H.O.Peitgen, D. Saupe:*The Science of Fractal Images*, Springer 1988

[PIE86] L.Pietronero, E.Tosatti (editors) *Fractals in Physics*, Elsever Science Publishers B.V., 1986

[SHN83] B. Shneiderman: *Direct Manipulation: A Step beyond Programming Languages* , in: IEEE Computer, 16, 1983, No.8, pp.57-69

[VOS75] R.F. Voss and J. Clarke:*1/f Noise in Music and Speech*, Nature 258 (1975) pp 317-318

[WAT88] Watkinson, John: *The Art of Digital Audio*, Focal Press, London and Boston 1988

Fractal Geometry in Vaporisation

J. H. Spurk
Technische Strömungslehre
Technische Hochschule Darmstadt

1 Introduction

In the liquid filled gap between two parallel plates, substantial negativ pressures can be generated, when the plates are separated with finite velocity. In untreated liquids, there are usually sufficient nuclei to give rise to cavitation bubbles, which will grow primarily by vaporisation from the cavity wall. The rate of growth will be limited by the reaction of the surrounding fluid, i.e. the inertial and/or frictional resistance of the fluid, and the ability of the liquid to provide, by conduction, the necessary heat of evaporisation to the bubble. For a bubble not in thermodynamic equilibrium (where the pressure in the bubble is not the saturation pressure at the temperature of the cavity wall) a rate equation will be required in addition to the effects mentioned.

Schneider [SCH89] has computed the growth under these conditions, even for the case of non equilibrium. But unlike the spherical bubble in an infinite fluid, the cylindrical bubble is not stable, as will be shown later.

The problem described, finds a practical realisation in the valve seat of fluid injectors for automobile engines, where the separation of the valve needle from its seat may lead to cavitation of the fuel, which subsequently causes faulty fuel metering to the engine. Schulz [SCH90] experimentally investigated the growth of bubbles in n-heptan, a common gasoline-like test liquid, in order to establish the bubble size, for which the bubble would still be cylindrical, a question of practical importance. His results show an almost immediate instability, and and a growth of the cavitation bubble in form of viscous fingers that appear to exhibit fractal geometry. Zeh & Eschenbrenner [ZEH91] have repeated the experiments for gasoline and methanol, a prospective automotive fuel, and again find a fractal growth pattern, but cylindrical bubbles at the initial stages.

It is the purpose of this paper, to provide an account of this work and give a stability analysis for a cylindrical bubble in the liquid filled gap when the plates are separated.

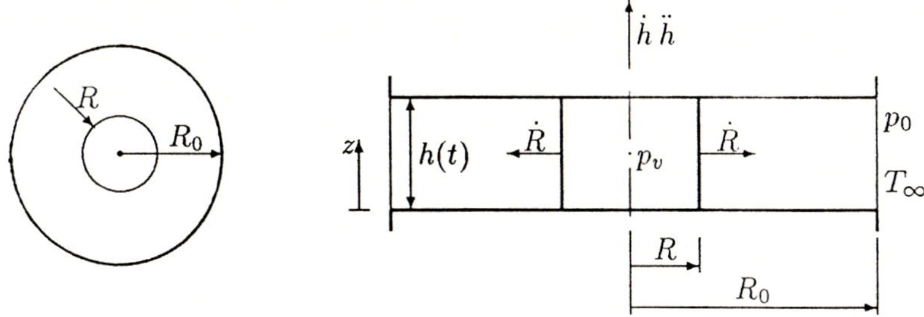

Figure 1: Sketch of bubble model

2 Theory

We shall consider the geometry shown in Fig. 1 and restrict ourselves to the simplest case i.e. equilibrium thermodynamics and infinite heat conduction. The pressure in the bubble is then the saturation pressure at the homogeneous liquid temperature. Under these conditions, the bubble will be driven by the pressure difference in the neighbourhood of the cavity and will grow in the pressure environment that is impressed by the friction and the motion of the upper plate. The problem then has a strong resemblance to the radial fingering in a Hele Shaw cell, reported by Paterson [PAT81], especially, if we consider first the case where $\dot{h} = 0$ and the pressure $p_0 < p_s(T_\infty) = p_v$. In fact, the simplifications introduced by Schneider are essentially the same as those leading to the Hele Shaw flow. We have for the average velocity $\vec{U}$ in the cell the equation of motion

$$\vec{U} = -M\,\nabla p \tag{1}$$

where

$$M = \frac{h^2}{12\mu} \tag{2}$$

and h is the spacing of the plates while μ is the viscosity of the fluid. (1) shows that $\vec{U}$ has a potential Φ, and for incompressible flow

$$\nabla^2\Phi = 0\ . \tag{3}$$

The solution of (3) statisfying the boundary condition

$$\nabla\Phi(R) = \dot{R}\,\vec{e}_r \tag{4}$$

is given by

$$\Phi = R\dot{R}\ln\frac{r}{R}\ , \tag{5}$$

and from (1) the pressure difference is

$$p_v - p_0 = \frac{\dot{R}R}{M}\ln\frac{R_0}{R}\ , \tag{6}$$

which provides the first order differential equation of the bubble radius. The solution is easily written down, but for $R \to 0$

$$\dot{R} \sim \frac{1}{R} , \tag{7}$$

so that inertial effects may not be neglected as R is very small. In the viscous fingering, the bubble is driven by a constant volume flux and therefore

$$\dot{R}R = const \tag{8}$$

and we have the same singularity, which however is not considered in Patersons work; in fact the radius can never be smaller than the injection hole for the volume flux. Here the radius can be very small and we may account for the inertia effect by introducing a volume force $\varrho\vec{k} = -\varrho\,\mathrm{D}\vec{U}/\mathrm{D}t$ into equation (1). (The gravitational volume force $\varrho\vec{g}$ is customarily introduced in linear fingering, which is of interest in oil industry where fingering reduces the amount of recoverable oil.)

However, the characteristic time scale is $\tau = \varrho h^2/(12\eta)$, which is the diffusion time across h. For $t < \tau$, inertia is important and for $t > \tau$ the inertia terms are negligible. For the condition of the experiments to be described later $\tau \sim 10^{-5}$ s and inertia would play a role only at times where the radius is very small and surface tension then stabilizes the bubble. This is confirmed by actual computation. In the case of separation of plates ($\dot{h} \neq 0$) the potential Φ no longer fulfills Laplace's equation, for from continuity we have

$$\nabla \cdot (h\vec{U}) + \dot{h} = 0 \tag{9}$$

and Φ now satisfies a Poisson equation

$$\nabla \cdot \vec{U} = \Delta\Phi = -\dot{h}/h \, . \tag{10}$$

The solution of (10) subject to the boundary condition (4) is

$$\Phi = R\dot{R}\ln\frac{r}{R} + \frac{1}{2}\frac{\dot{h}}{h}\left(-\frac{r^2}{2} + R^2\ln\frac{r}{R}\right) . \tag{11}$$

This solution is valid irrespective of inertia being considered or not. When (11) is introduced into equation (1), augmented by the volume force $-\varrho\mathrm{D}\vec{U}/\mathrm{D}t$, which has a potential, we find a second order differential equation for the growth of the bubble. The solution starts from the initial conditions $\dot{R}(0) = 0$, $R = R_i$, where R_i is chosen such that the cylindrical bubble has the same volume as the equilibrium spherical bubble. A numerical solution will be shown later. The solution however is not stable with respect to disturbances of the bubble radius r_v

$$\eta(\varphi, t) = r_v(\varphi, t) - R(t) = \eta_0\,\mathrm{e}^{\beta t + in\varphi} \tag{12}$$

with $\eta_0/R \ll 1$. Therefore, if we put

$$\Phi(r, \varphi, t) = \Phi^{(0)}(r, t) + \eta_0\,\Phi^{(1)}(r, \varphi, t) \tag{13}$$

we obtain from (10)

$$\Delta\Phi^{(1)} = 0 \tag{14}$$

and the solution which satisfies (4) and vanishes as $r \to \infty$ is

$$\Phi^{(1)} = -\frac{R}{n}\left(\beta + \frac{\dot{R}}{R} + \frac{\dot{h}}{h}\right)\left(\frac{R}{r}\right)^{n} e^{\beta t + in\varphi} , \tag{15}$$

where $\dot{h}$ is a given function and $R(t)$ follows from the solution of the differential equation mentioned above.

Continuity of normal stress across the cavity wall, neglecting viscous stresses, leads to

$$p^{(0)}(R+\eta) + \eta_0 p^{(1)}(R+\eta) = p_v - \sigma\kappa \tag{16}$$

where σ is the surface tension and κ is the curvature. This boundary condition determines the rate of growth β.

If inertia terms are neglected we find

$$\beta = \frac{n-1}{R^3}\left(\dot{R}R^2 - n(n+1)\sigma M\right) - \frac{\dot{h}}{h} , \tag{17}$$

which reduces to Paterson's result [PAT81] for $\dot{h} \to 0$, i.e. to the radial fingering.

The wavenumber n having maximum growth β follows from $\partial\beta/\partial n = 0$ and is independent of $h(t)$, so the instability follows the route of viscous fingering.

3 Apparatus

The liquid film is formed between the gap of two parallel glass plates having a diameter of 25 mm. The plates are flat to 1/10 of the wavelength of He-Ne laser light. The upper plate is mounted in a precision mount that can be tilted by way of three micrometer screws. The lower plate is attached to a spring support which acts as a parallel guide, keeping the face of the lower plate allways parallel to the upper plate during separation. This spring mount is held under pretension by a wire. Upon melting this wire by a current pulse, the lower plate moves away from the upper, generating negative pressures in the liquid. This motion is recorded, and at preset time during the motion, photographs of the bubble patterns are taken with exposure times of less than 20 nano s.

The gap is adjusted before the test to gap height between 5 to 10 μm and parallelism of the plates is checked by oberserving the interference fringes from the gap in He-Ne laser light.

4 Experiments

First the motion of the mount is determined by solving the equation of motion under the pretension and the pressure force on the plate. From this solution $h(t)$,

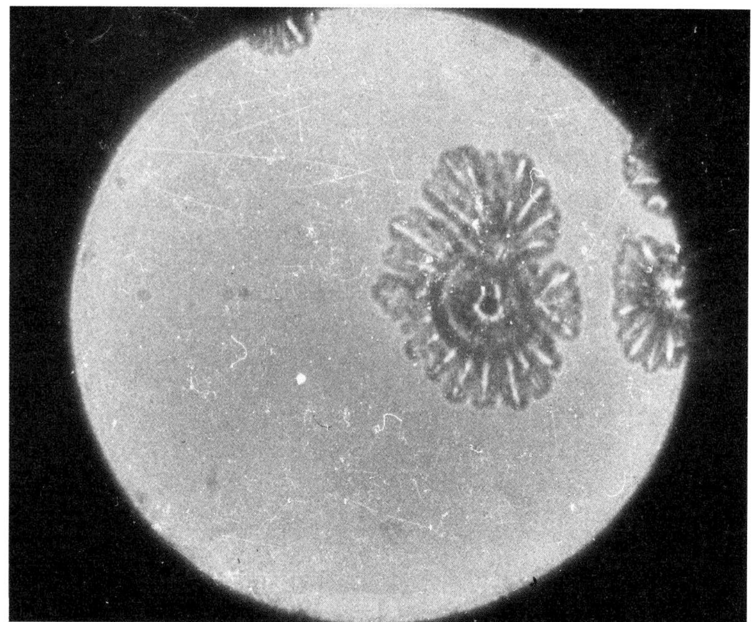

Figure 5: n-Heptan $p_s(20^0) = 0.047$ bar, $\Delta t = 0.5$ ms

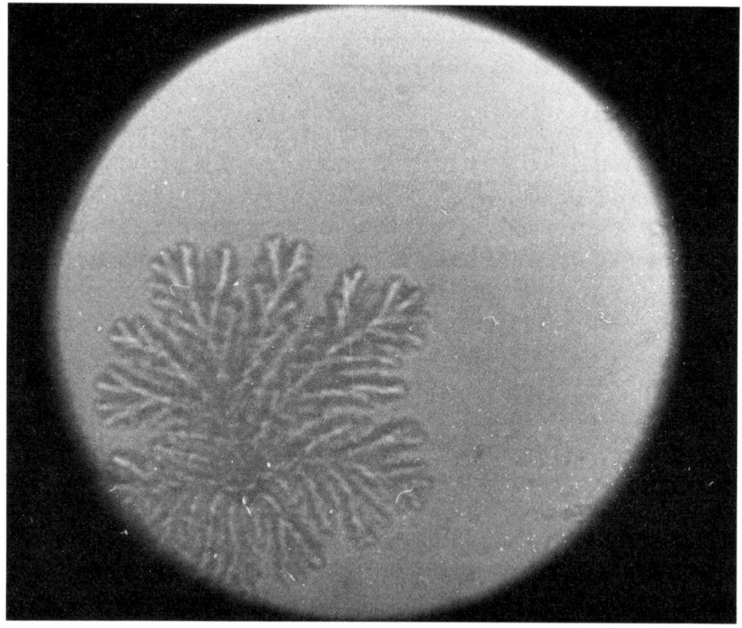

Figure 6: n-Heptan $p_s(20^0) = 0.047$ bar, $\Delta t = 3.2$ ms

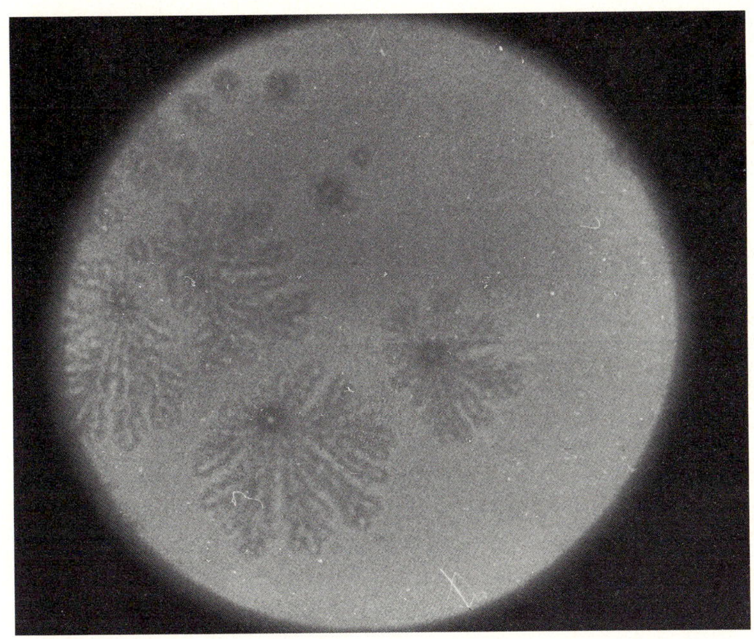

Figure 7: Methanol $p_s(20^0) = 0.13$ bar, $\Delta t = 3.0$ ms

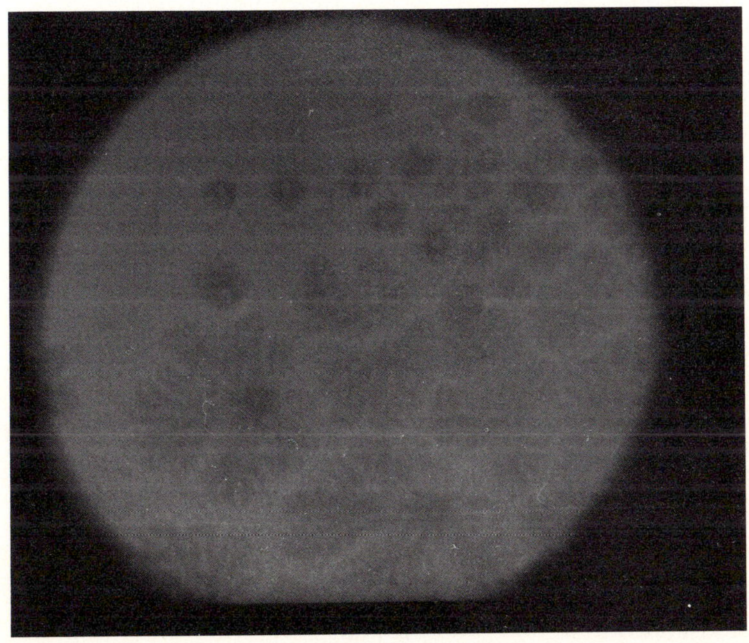

Figure 8: Gasoline, $\Delta t = 2.2$ ms

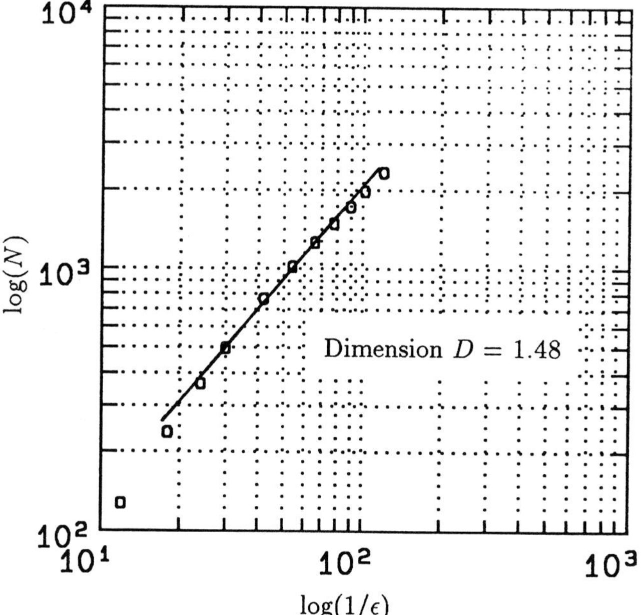

Figure 9: Fractal dimension of bubble contour, n-Heptan

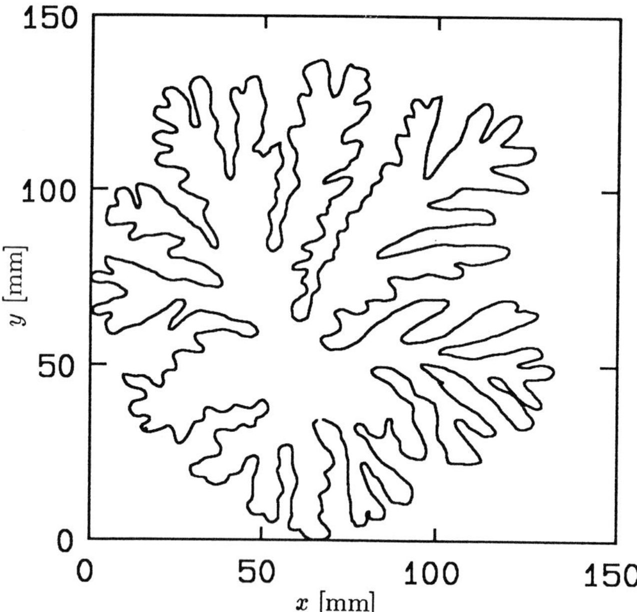

Figure 10: Bubble contour, n-Heptan

wire material was probable. When contamination was avoided and under clean test conditions, many tests using n-heptan showed no bubble at all. n-heptan has the lowest saturation pressure but the pressure in the gap center was always negative during tests. Our experience shows that in methanol usually more than one bubble is formed. Here almost cylindrical bubbles can be observed. Gasoline evaporates much earlier and readily and a number of cylindrical bubbles can be seen. In all pictures however fractal patterns appear to be present. The fractal dimension is not easily extracted from these pictures since the boundaries are not sharply seen. All the liquids have a rather high index of refraction which affects the spatial resolution.

To compute the dimensions, the boundaries were drawn on enlarged photographs and these where used in the data reduction. Fig. 9 shows the usual plot of $\log(N)$ versus $\log(1/\epsilon)$ where N is the number of meshes covering the curve and ϵ the mesh size for the bubble shown in Fig. 10. As the mesh size is smaller than the lowest scale of the bubble, the dimension $D = 1$ is reached.

References

[PAT81] L. Paterson: Radial fingering in a Hele Shaw cell, J. Fluid Mech. (1981), Vol. 113, pp 513-529

[SCH89] J. Schneider: Berechnung der Kavitation im Schmierspalt unter Berücksichtigung endlicher Verdampfungsraten, Diplomarbeit Fachgebiet Technische Strömungslehre, Technische Hochschule Darmstadt 1989, unpublished.

[SCH90] R. Schulz: Versuchsstand zur photographischen Untersuchung von Kavitationsblasen zwischen zwei sich auseinanderbewegenden parallelen Platten, Studienarbeit, Fachgebiet Technische Strömungslehre, Technische Hochschule Darmstadt 1990, unpublished.

[ZEH91] F. Zeh, J. Eschenbrenner: Experimentelle Untersuchung zum Fraktalen Wachstum von Kavitationsblasen, Studienarbeit, Fachgebiet Technische Strömungslehre, Technische Hochschule Darmstadt 1991, unpublished.

MRT Imaging of Time Dependent Processes

J. Syha and A. Haase

Physikalisches Institut, Universität Würzburg

1 Introduction

The importance of MRT for clinical and biological applications has increased rapidly in the last decade. Improved hardware of MR tomographs together with new imaging techniques have lead to a wide use in many investigations. In the last few years research has focused especially on fast MR and on high resolution (MR microscopy) imaging techniques.

Among the reasons to reduce the scan time, economics and patient comfort are the most significant. In comparison to radiographic computed tomography MRT is relatively slow with scan times of several minutes, the number of patients is restricted and the cost relatively high. Beside of cost and patient comfort speed enhanced MRT has an other important implication, the increase of possible investigations with MRT. Successful abdomen imaging with conventional MRT is disturbed by respiratory and peristaltic motions. Scan times as small as a fraction of a second will freeze this motions and even allow MRT of the heart. Additionally to the fact that fast imaging techniques allow the acquisition of distortion free images of moving parts of the body they also permit examinations of the motion itself. The power of fast scan techniques in MRT is far beyond the scope of its use until today and many new applications will be invented in the near future.

In this article we will explain advantages and limitations of fast MRT, based on NMR parameters, and additionally present some examples of possible applications of this new method.

2 Fast Scan MRT: Advantages and Limitations

Signal intensity in MRT depends on many parameters, i.e. the type of nucleus used for the investigation, the concentration of this nucleus within the object, the relaxation times, flow and diffusion of the exited spins during the experiment, movements of the object, etc.. To achieve the appropriate conditions for a MR experiment in respect to optimized signal intensity, the operator has to consider the influence of these parameters.

The choice of the nucleus depents first of all on the interest of the investigator, the process or status he likes to examine, and is strongly limited by the NMR sensitivity of the nuclei and by their concentration within the object. The nuclei which are of interest for biomedical applications are ^{1}H, ^{19}F, ^{23}Na, ^{31}P, ^{35}Cl and ^{13}C. Because NMR sensitivity is given by the gyromagnetic ratio and natural abundance of the nucleus, ^{1}H is the most sensitive nucleus followed by ^{19}F and the other nuclei mentioned above. Furthermore, the concentration of ^{1}H is approximately 100 mM, three to six magnitudes higher than all other nuclei, because water is the most prominent substance in biological tissues. With fast MRT methods the problem of sensitivity is pronounced, as speed is increased on the cost of signal, and therefore fast MRT is restricted to protons.

The second highest signal in proton MRT derives from the methylen groups of fatty acids in adipose tissue. Due to the different chemical environments of fatty acid and water protons their Lamor frequences differ slightly, known as chemical shift. Using high field strenght (> 2T) and high pixel resolution it can leed to artifacts in the MRT because each resonance gives an image of its own and is slightly shifted against the other. Furtunately, in clinical applications with field strength up to 1.5 T the chemical shift difference between water and fat protons does not reach the pixel resolution. Although in usual clinical MRT one does not have to concern with chemical shift artifacts, there are applications which depend on or at least gain information from it. It is possible to acquire selective images from water and fat protons by using appropriate pulse sequences. For investigations of lipid tumors and of bone marrow this is a very useful feature and it can be carried out using both, conventional and fast scan MRT techniques.

Contrast in clinical MRT is primarily gained by the difference in relaxation times of the spins in different tissues. T_1 or spin-lattice relaxation time is the time constant of the return of the longitudinal magnetization and T_2 or spin-spin relaxation time the time constant of the return of transversal magnetization of the spins to equilibrium. Contrast in the images can easily be changed by variation of experimental parameters, which result in different T_1 and T_2 weightings. T_1 contrast depends on the pulse angle used for excitation and on the repetition time (TR) of the pulses. T_2 contrast can be varied by changing the delay between the pulse and the acquisitions. Contrast in conventional MRT is always a mixture of both, T_1 and T_2 effects, only the strength of their influence can be changed so that one is pronounced. It is common use to take two images, which were acquired with just one different experimental parameter, to estimate a relaxation time. For example, the difference in signal intensity after variation of the TR-delay will purely depend on T_1 relaxation. Relaxation time values calculated from such images are inaccurate and they vary widely between different systems. With fast scan techniques several images can be acquired within a second and therefore 10 to 20 images are easily produced. Using the fast scan technique relaxation times can then be calculated on the basis of a big number of images as accurate as spectroscopically measured relaxation times.

Flow, diffusion and movements all decrease the signal in MRT and can produce severe artifacts. Techniques have been developed to obtain images only from flowing liquids, i.e. blood, thereby producing images only of blood vessles. Heart and breathing motions can be avoided by appropriate triggering, but then the scan time is even longer. Artifacts of peristaltic and other irregular motions can not be avoided by any conventional MRT technique. Therefore the most popular organ for investigations with MRT is the brain.

Fast imaging techniques have poor sensitivity but they show less artifacts due to motion, diffusion, and flow. It is even possible to study motions and observe them in movies, which can be produced by linking of the images. Using special pulse techniques parameter weighted images can be performed as in conventional MRT and, what is even more important, quantitation of NMR parameters can be achieved faster and more accurate. The technique is even fast enough to follow up the dynamic changes of these parameters after the application of a substance effecting them. For example, after application of a contrast agent the changes in T_1 relaxation times can be monitored with a resolution of 3 seconds. Thereby it is possible to follow up the time course of the distribution of the agent within the body.

3 Principle of fast MRT

The fastes MRT technique, called Echo Planar imaging (EPI), with measuring times between 32 and 128 ms has been invented by Mansfield 15 years ago [MAN77]. Unfortunatelly this technique can only be applied on systems with special hardware configurations, i.e. grandient coils with a switching time as short as a fraction of a millisecond. Until recently such gradient coil systems were not offered commerically. Therefore other fast techniques with measuring times in the order of a few seconds have been introduced, like the Fast Low Angle Shot (FLASH) technique [HAA86]. Improvement of the gradient coil systems in the last few years not only made EPI available for clinical examinations but also allowed to decrease the imaging time of FLASH as in snapshot FLASH [HAA90]. Snapshot FLASH images can be performed within 340 ms with a 128 x 128 pixel resolution, which still is about 10 times slower than EPI.

EPI and snapshot FLASH are two completely different MRT techniques. While EPI can be described a a single-shot MRI, using one exitation of the selected plane which is followed by a series of MR signals created by a frequent reversal of magnetic field gradients, snapshot FLASH is a multi-shot technique, where many exitation pulses are needed to acquire an image. According to this, both techniques have their advantages and disadvantages. For example signal strength with EPI is 10 times higher than with snapshot FLASH because a 90° pulse is applied, while snapshot FLASH needs excitation flip angles of less than 5°. On the other hand using snapshot FLASH a higher spacial resolution can be achieved, due to the fact that for higher resolutions the number of acquired echos have to be increased and that of course effects the scan time in both methodes. But with EPI, where only one excitation pulse is used for the whole sequence, the intensity of the echos decays due to T_{2^*} relaxation and restricts the number of possible acquisitions.

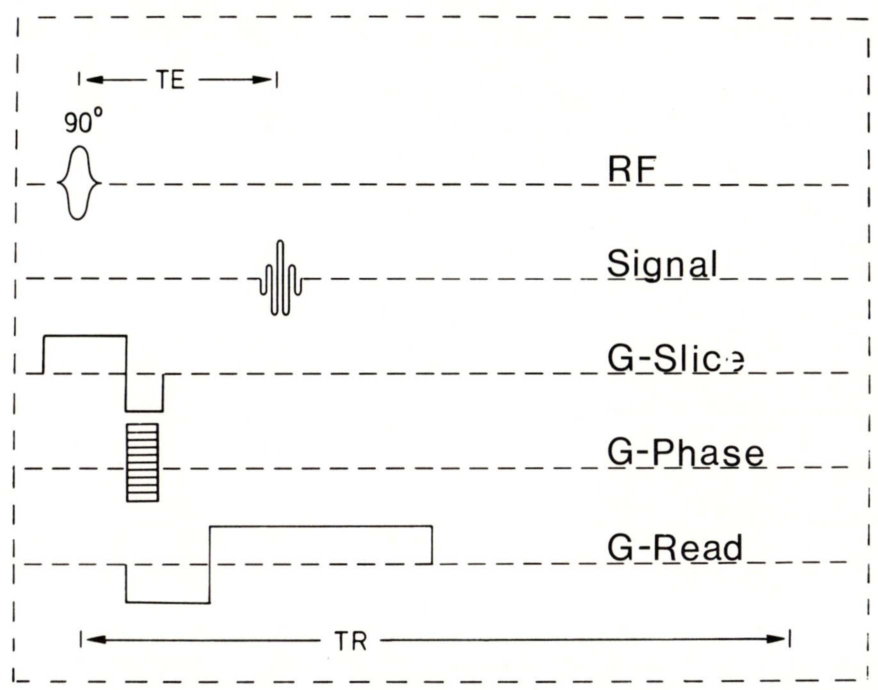

Fig. 1: Timing diagram for snapshot FLASH

The snapshot FLASH technique we use is illustrated in Figure 1. It can be regarded as a member of the family of gradient-echo pulse sequences, as the echo is produced by the inversion of the read gradient before the acquisition. Additionally, performance of the acquisition of the signal in the presence of this gradient yields information about the distribution of spins across this gradient. To perform slice selection a slice gradient is switched on during the application of the radiofrequence pulse. A phase encoding gradient is used to gain information in the second dimension of the image. All gradients are orientated orthogonal to each other and depending on which gradient is used for slice selection, transaxial, saggital or coronal images of the object can be obtained. We perform images of a 128 x 128 pixel matrix within 340 ms by using a TR of 2.6 ms and a TE of 1.4 ms.

A snapshot FLASH image represents the spin density of the protons and therefore shows poor image contrast. In general T_1 and T_2 contrast could be obained using longer TR or TE delays, respectively, but this would again prolong scan times. Superior to this is the use of an appropriate preparation pulse sequences prior the performance of the snapshot FLASH sequence, which allows to obtain any desired contrast. The simplest preparation is the application of an inversion pulse, which provides different T_1 contrasts in the images depending on the delay between the inversion pulse and the snapshot FLASH sequence.

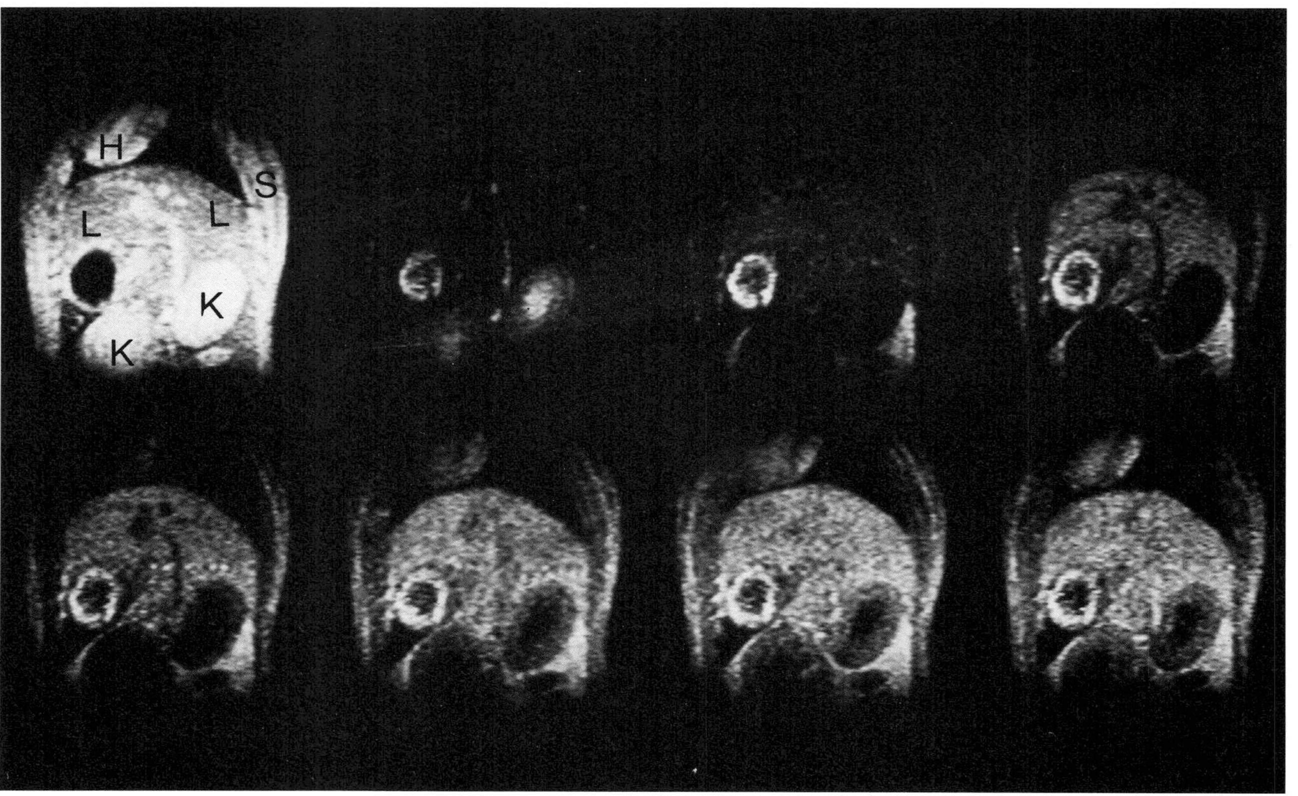

Fig. 2: T1 weighted snapshot FLASH MRT of rat abdomen. The 8 images had been scanned after one inverion pulse and the t1 contrast changes from image to image
H = heart, K = kidney, L = liver, S = chest wall, including skelettal muscle

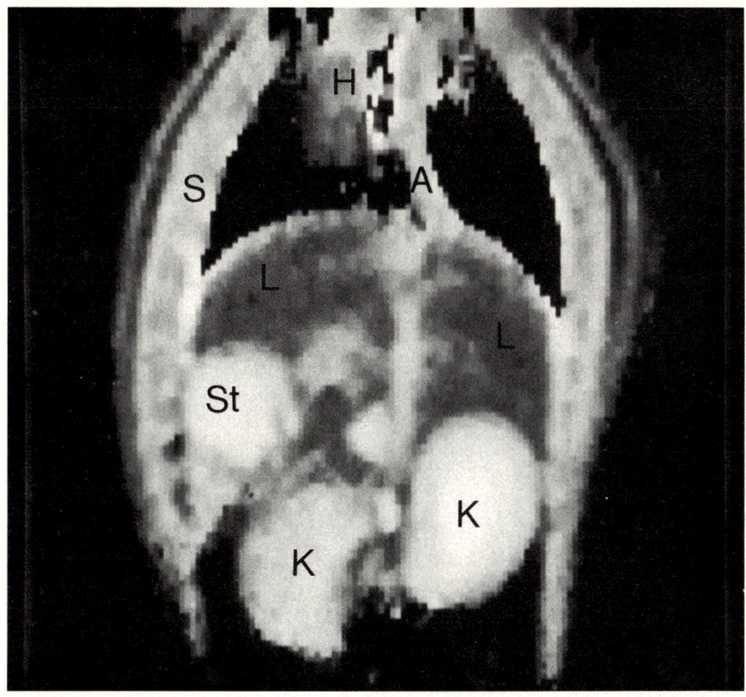

Fig. 3: T_1-map of rat abdomen
A = aorta, H = heart, K = kidney, L = liver, S = skelettal muscle, St = stomach

4 Examples of fast MRT

In Figure 2 diffent T_1 contrasts are demonstrated on a coronal plane of rat abdomen. All 8 images were taken after the application of one inversion pulse. The T_1 contrast changes from image to image due to the increasing delay between the inversion pulse and the image acquisition. Although the signal intensity is poor compared to conventional imaging techniques the images contain no motion artifacts. This experiment demonstrat the power of the technique. It can be used in many varieties of which we like to show only two.

Most of the contrast agents used with MRT decrease the T_1 relaxation time of the spins. They are used to increase the contrast between pathological and healthy tissues in conventional clinical MRT examinations. We used contrast agents in rats to observe their distribution dynamically. For this purpose we used the snapshot FLASH technique with an inversion pulse prior each image. Maximal changes in signal intensities after the application of the contrast agent is gained when the signal intensity of the images is minimal in the pre-contrast image. This can be achieved by carefully adjusting the delay between the inversion pulse and the acquision of the image. We used this method

of the contrast agent the T_1 relaxation time decreases from 2.5 to 0.2 seconds in kidney medulla and this low value is maintained over 6 minutes. Then a slight increase is observed to about 0.5 seconds after 20 minutes. In liver the minimal T_1 relaxation is reached already after 30 seconds and this low level maintained for at least 20 minutes.

Studies of motions is another important application, which can be performed easily. Even 3D studies of motion are possible within a few minutes [HEN90], hence a 3D experiment can be completed within 32 heart beats.

5 Literature

[DEI92] R. Deichmann, A. Haase: J. Magn. Reson., in press, expected Feb. 1992

[HAA86] A. Haase, J. Frahm, D. Matthaei, K.D. Merboldt, W. Hänicke: J. Magn. Reson. 67, 258, 1986

[HAA90] A. Haase: Magn. Reson. Med. 13, 77, 1990

[HEN90] D. Henrich, A. Haase, D. Matthaei, Magn. Reson. Imag. 8, 377, 1990

[MAN77] P. Mansfield, J. Phys. C10, L55, 1977

[NEK91] S. Nekolla, T. Gneiting, J. Syha, A. Haase: J. Comp. Assis. Tomogr., in press, expected Nov. 1991

[SYH90] J. Syha, R. Bartkowski, F. Cavagna, A. Haase, D. Henrich, S. Shadam, Ninth anual meeting of the Society of Magnetic Resonance in Medicine, New York, Abstract 55, 1990

II. Part: Computer Graphics

Fractals and Formal Texture Specification

Gabriele Englert, Manfred Schendel

Fraunhofer–Institut für Graphische Datenverarbeitung, Darmstadt

As a rule, the surfaces of real objects are not structureless. In addition to a constant color they have further properties that are commonly summarized under the term texture. Textures are location–dependent variations in color, geometry, or even transparency or surface roughness, and give the characteristic appearance to materials like stone, sand, marble, and textile surfaces.

Thus, besides the modelling of pure object geometry, the presentation of surface properties is decisive to the quality of computer–generated images of realistic scenes. The traditional processing method for this application of textures is to compute a two–dimensional image array of the texture (*texture synthesis*) and to coat it on the corresponding object surfaces (*texture mapping*).

Fractal algorithms play an increasing role in the process of texture synthesis (e.g. [Ba*etal*88], [DeHN85], [Falc85], [Gard85], [HaBa84], [Krue88], [LaKa87], [PeSa88], [Pent84], [Smit84], [Upso86]). They are especially applicable for the generation of stone–like surfaces, marble structures, or cloud formations. This is surprising, since fractals are usually conceived as fractal geometry. Object modelling in computer graphics, however, distinguishes rigorously between pure object geometry and surface texture: Object geometry consists of edges, points and surfaces; textures are arbitrary structures in raster format that are mapped to the object's geometry. In this way, any image can be used as surface texture. Integrated texture synthesis systems, however, use formal specification methods with semantic informations ([Kauf88], [WMN85]). Therefore, the application of fractal algorithms demands the consideration of the semantic conditions.

Texture specification and synthesis systems are required because the synthesis algorithms are based on complex mathematics and, thus, a controlled generation of textures is usually only gained by people with mathematical or technical educations. The users of textures, however, come from artistic professions like architecture, industrial design, or movie animation. Therefore, an integrated tool is demanded for specification, generation and testing of textures with predefined geometric objects. In addition, already realized textures have to be managed and archived [KöBE90].

It is obvious that the development of such systems needs more than an attentive system design. First of all, a consistent interpretation and, thus, a comprehensive definition

rectly in the scene description, whereas non–constant surface properties are represented by pointers to externally stored discrete value samples (picture 2) .

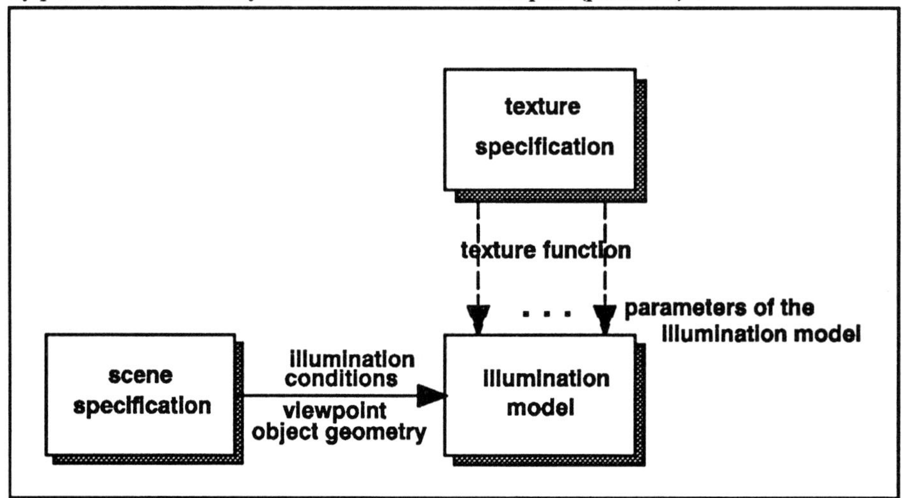

Picture 2: Texture functions as the input of the applied illumination model

During the visualization process (*rendering process*), the evaluation of the surface properties is carried out in the illumination calculation, that simulates the light–material–interaction. The physical phenomena like color, transparency, surface roughness, etc. are processed as independent quantities.

Hence, texture sensations are regarded and processed as seperated and independent values. Therefore, in computer graphics the comprehension of the term "texture" differs from the definition of texture given by the perception physiology. The term *texture* stands here for the textured structure producing the texture sensation. Moreover, textures are not restricted to two dimensions, but comprise three dimensional or / and time–variant structures like e.g. clouds [EnSa89].

3 A Reference Model for the Description of n–Dimensional Textures

Following the previous chapter, a reference model for textured surfaces describes the structures of location–variant functions, that are used as input parameters of the illumination model.

A *complete texture function* that is handed directly to the illumination model, influences all of the corresponding parameters. It consists of several layers, corresponding to the input functions of the individual parameters of the illumination model. The layers of a texture may be correlated. Eventually, some layers are specified as constant values.

The construction of texture functions starts with so–called *generating functions*, comprise placing rules of texture elements, too. By means of monadic operations like filters on color values or combination like the overlay function, a hierarchical structure is built up, which is represented by an acyclic graph (*structure graph*) (see picture 3).

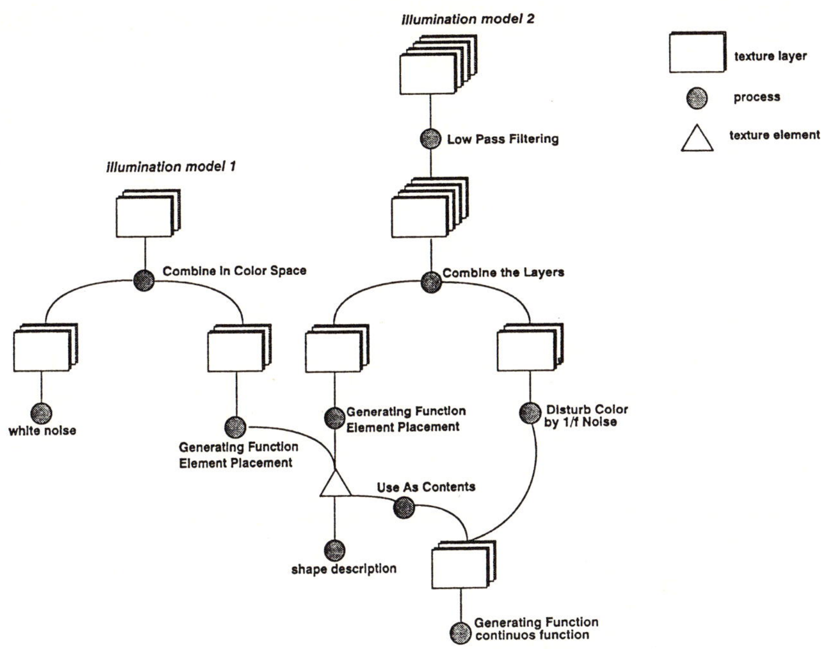

Picture 3: An example for the hierarchical structure of texture

The following description of the texture reference model concerns after a short survey only those topics which are of interest for the integration of fractal structures. Any formalism is omitted. For a more detailed discussion please see [EnSa89] and [E*etal*90]..

3.1 The Construction of Texture Functions

For the construction of texture functions, the texture reference model provides the following object types:

- *Domains of texture elements*
  A domain consists of its boundary specifying the shape and the size of a texture element, and of the set of its inner points, which, at the end, represents the domain of the texture element. The boundary of the domain must be a closed path, possibly composed of a finite number of continuous curves.
- *Texture elements*
  A texture element is described by its domain and its contents; the latter are complete texture functions. Both of them are defined in a joint local coordinate system.
- *Textures*
  As mentioned above, textures are described as location–dependent functions. In or-

with *texture types* matched to the textures or elements. While specifying the texture function for a particular layer, one has to consider that the individual layers demand specific data types. *HiLDTe* supplies several usually required layers automatically, but it is also possible to add new application–dependent layers. At present, *HiLDTe* provides the following layers per default:

layer	data type
transparency	percent
reflection	percent
material refraction	float
surface roughness	float
surface normal	vector
colour	abstr_color

In order to reproduce the hierarchical structure of textures and to simultaneously provide an understandable programming surface, *HiLDTe* is based on a block concept, in which definition procedures are interpreted as objects, which may contain subobjects recursively. The range of validity of naming conventions, defined objects, texture types, etc. reminds of PASCAL: Names as well as objects are locally known in that block in which they have been declared, and they are globally known in all subobjects of the actual block.

The logic of the texture model implies some restrictrions to the hierarchical program construction, e.g., it is not allowed to specify a texture in an element domain.

Processes as well as definition procedures can be recorded in application libraries and then used in each block as *externals.*

4.1 *HiLDTe* Components

The definition procedures are the principal items of a texture description in *HiLDTe*. The result of a definition procedure is always an independent module.

4.1.1 Element Domains

The specification of element domains consists of the description of the shape and the size, both relating to the local coordinate system.
For the shape definition, following constructs are available:

- basic geometric forms like circle, ellipse, rectangle, symmmetric polygons
- closed pathes consisting of lines, splines, mixtures of lines and splines
- combinations of two domains by means of logic operations like difference, union, and cut
- deformations of an existing domain with the help of stochasting processes that act on the domain's boundary.

The following simple example illustrates the syntax of domain specification:

```
DOMAIN circle (rad: float);
BEGIN
   BOUNDARY =CIRCLE (rad)
END.
```

4 . 1 .2 Texture Elements

Elements are made up by their shape and size (see the above description of element domains) and by their contents which are part of an arbitrary image or texture function. As mentioned above, elements have a texture type.

Example:

```
ELEMENT blue_circle (rad: float) : COLOR;
BEGIN
    BOUNDARY            = CIRCLE (rad);
    CONTENTS .COLOR = ACNS (BLUE)
END.
```

4 . 1 .3 Processes

Processes are mainly used for instantiating specified objects, especially for parameterisizing and placing elements into the definition space. They are also applied for the generation of pseudorandom numbers for texture manipulations and provide facilities of the G1–texture functions (see chapter 3.2).

According to the task of producing tupels of numbers, processes are the only *HiLDTe* procedures that have output parameters. The output is managed clockwise with the help of construct "OUTPUT".

HiLDTe supplies a set of standard processes, that may be used to build more complex and more special application processes. Examples are GRID, INTERVAL, INTERVAL_2D, BINOMIAL, NORMAL, NORMAL_2D, etc.

Example:

```
PROCESS elem–pos –> (pos:POSITION);
var cnt: INTEGER;
FOR cnt DO BEGIN
      pos = INTERVAL2D (0,1);
      OUTPUT
END
```

4 . 1 .4 Textures

Specifying textures means to specify the contents of their layers.

While G1–functions (see description of the texture model) are matched directly to the layer, G2–functions use the function "SCATTER", that combines the element description with the instantiating and the placing process.

The following example illustrates both a G1–texture (as background) and a G2–texture (as foreground).

```
TEXTURE easy : COLOR;

ELEMENT elem–name (pos: POSITION, rad: FLOAT) : COLOR;
BEGIN
    BOUNDARY = TRANSLATE (pos, circle (rad));
    CONTENTS = ACNS (BLUE);
END;

PROCESS elem–proc –> (pos: POSITION, rad: FLOAT)
var cnt: Integer
FOR cnt DO BEGIN
    rad  = 0.03
    pos  = INTERVAL_2D (0, 1)
    OUTPUT
END ;

BEGIN
    CONTENTS.COLOR  = RGB (INTERVAL(0,1), INTERVAL(0,1),
                                    INTERVAL(0,1));
    CONTENTS.COLOR  = SCATTER (elem_name, elem_proc);
END;
```

4.2 Adding new Facilities to HiLDTe

There exist two main possibilities to add new features to the language HiLDTe. One is the integration of new processes, the other is the definition of special G1–functions and, thus, embedding the new feature as a "texture". In contrast to processes, those textures may only be used in the assignment

```
CONTENTS.layer = texturename ( ... );
```

This solution is only efficient, if the feature operates directly on special texture layers, i.e., the result of the feature is an array with values of the data type according to the specified layer.

5 Fractal and Texture

5.1 Usability of Fractals for Texture Generation

Texture synthesis by means of fractal algorithms is carried out in two steps: The first step produces a fractal data record, using one of the well–known procedures like boundary scanning method for Julia sets or the spectral synthesis ([Mand82], [PeSa88]).

The second and mostly the more difficult step maps the fractal data record onto the texture space. By doing that, further computation may transform the data record into an adequate representation. One of the important tasks of the second step is the projection of the arbitrary dimensional data record into the dimension of the texture space.

The usual application of fractal algorithms in texture synthesis corresponds to the way ordinary texture generation is performed. A data array with fixed resolution is produced for each individual texture layer -- and, thus, possesses all disadvantages of the straight–forward texture generation, we mentioned above. Additionally, the parameterization of fractal algorithms cannot easily be understood.

This problem may be solved by a capable texture editor supplying a shell for parameter translation. The system user will then work with values like orientation, granularity, contrast, etc. Thus, non–experienced users have a chance to apply fractals in texture synthesis in an efficient manner.

An integration of fractal models into a texture synthesis language allows manifold interpretation of fractals, as colors, as height fields, etc. On the other hand fractals could be used in combination with such integration algorithms in order to provide new structures.

5.2 Classification of Fractal Construction Algorithms

Of course, a system supported integration into the texture synthesis language HiLDTe is not possible for each special fractal algorithm. However, with a small number of universal algorithms, a large number of fractal models could be made available. In order to find them it is necessary to discover common aspects of fractal construction algorithms and, thus, to classify them. Besides, the user may implement special algorithms as application dependent processes.

Given a fixed set of construction parameters, the classification is based on applied generation processes. We have found out four different classes that are presented in the following paragraphs:

- *Recursive Construction*

  Algorithms of this class construct the model of the fractal recursively. In each step, the parts of the fractal model are subdivided into more pieces. The fractal model could be assigned to a generating function of type G2b of the texture model. The most important common aspects of these algorithms are, that in each calculated step the details of the model are increased exponentially. But of course the processing time increases, too.

  Famous examples of this type are algorithms based on midpoint subdivision. The main applications are the construction of mountains or clouds by means of Brownian motion. Another example is the evaluation of iterated function systems (IFS) by a multicopying machine.

  At high resolution, the results of the algorithms above could be mapped onto a plane. The achieved texture would be of type G1b.
- *Iterated Systems*

  The second–class algorithms are based on iterations. During each step, the intensity value of one point of the texture plane is calculated by an iteration. The most famous member of this class is the boundary scanning method for the calculation of Julia or Mandelbrot set representations.

  Any other two– or higher dimensional equation system could be used here. Usually, the speed of convergence is used as fractal information. Depending on the requested resolution, the plane is divided along the x– and y–axis into a corresponding number of fields. The midpoint of each field represents the starting value for the iteration. These construction algorithms could be used easily for all resolutions. The resulting textures are of type G1a.

– *Orbit Calculation Systems*

These algorithms are using only one iteration equation. For two–dimensional textures, two–dimensional or complex functions are used. In each step, the resulting value pair is mapped as one point onto the plane. Increasing the number of steps, an orbit becomes visible on the plane. The probability to set a new point depends on the number of iteration steps and on the resolution of the texture plane.

One important example of algorithm of this class is the calculation of the Hénon attractor. The resulting textures are of type G1b.

– *One step synthesis*

One famous member of this last class of fractals is the spectral synthesis of mountains or clouds based on a frequency spectrum derivated from Brownian motion. Here, a frequency spectrum is generated for a fixed resolution, and subsequently an inverse discrete Fourier transformation is performed. The resulting values could be mapped onto the plane by using different colors for different values. The result of this application reminds artificial maps or clouds depending on the used colors. To change the resolution of the fractal model, a complete recalculation is necessary. In small ranges for texture mapping it is possible to interpolate points between the given height field. The textures are of type G1b.

It is not the intention of this paper to describe some fractal construction algorithms. For more information about the algorithms, refer to e.g. [PeSa88], [Mand82].

Thus, we have four classes of fractal construction algorithms that are suited for texture synthesis and corresponds to texture classes of the introduced texture reference models.

5.3 The Integration of Fractal Algorithms into HiLDTe

The next step is to integrate these fractal generation algorithms into the texture synthesis language HiLDTe. As mentioned in the previous chapter, there are two different possibilities. The first is to produce a parameterized external TEXTURE object, which is directly specified for one or more texture layers. By doing this, the calculation of the intensity values relates strongly to the data type of the used texture layer, and there is no chance to change the assignment at the level of the texture synthesis language.

The usage of PROCESSes permits a more general application. By means of the varibale LOCATION, processes are able to describe location–dependent variation in the texture space. LOCATION is automatically defined in each definition procedure with the type *ELEMENT* or *TEXTURE* and possesses the data type POSITION. The assignment to one or more arbitrary texture layers can be completely performed at the level of the texture synthesis language HiLDTe. Moreover, a process may be used for a complex construction of process blocks.

The user who just wants to generate nice textures should not have to comprehend the mathematical fundamentals of fractal model generation. The proposed solution requires only knowledge about the value range of the wanted texture layer and the value range of the resulting intensity values. Of course, it is not easy to specify fitting parameters if the theoretical background is unknown. It is the task of a powerful and intelligent user interface to provide the support for the efficient handling of textures.

The integration of fractals in HiLDTe as time–discrete processes is demonstrated by means of the iteration system of Hénon's equation and the IFS–system.

5.4 An Example: The Hénon Iteration System

The equations of Hénon describe a chaotic dynamic system. It is well–known for esthetic pictures resulting from iterations with arbitrarily starting values. The pictures are generated by mapping the resulting pair (x, y) of each iteration onto the texture plane. The gained orbit is a strange attractor and, thus, it describes a fractal.

In the following example, we do not want to calculate the attractor but convergence aspects. The speed of convergence or divergence is mapped onto different colors for each starting value taken from the location of the point in the plane.

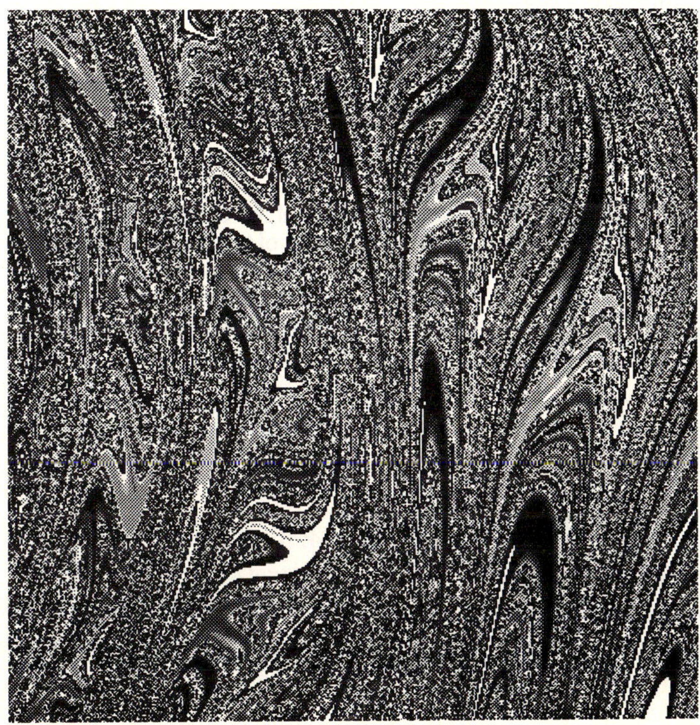

Picture 4: The visualization of an Hénon iteration system

The successive recursive steps on the shape of the letter "F" as texture element are shown in picture 6.

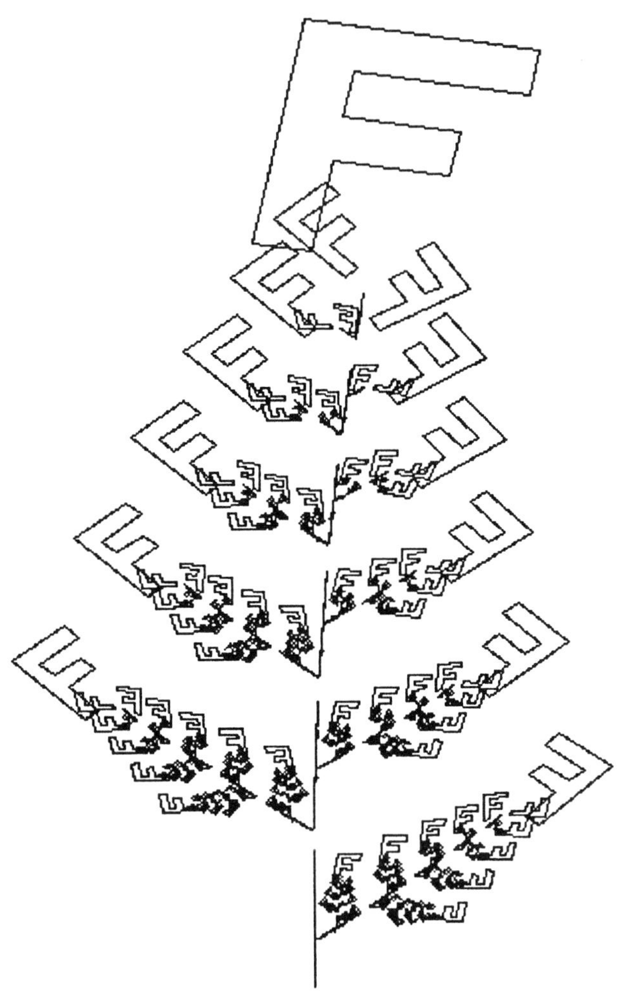

Picture 6: A non–texture example for IFS–code

A possible integration of IFS–code fractals is to use the G2–functions (see chapter 3.2) and to implement the algorithm as a *HiLDTe* process for placing and parameterizing texture elements. The following example will illustrate this application:

```
TEXTURE ifs–system : COLOR;

/*  Recursive placement and element changes */

EXTERNAL      PROCESS RECURSIVE;
```

```
/* specification of the process RECURSIVE:
 * RECURSIVE ( (ifscode) : LIST; nrecurs : INTEGER)
 *                 -> (pos: POSITION; rot: FLOAT; size: FLOAT)
*/
/* input parameters:
   ifscode        list of  transformations (ifs code )
   nrecurs        number of iterations

  output parameters:
   pos            position of the element
   rot            orientation of the element
   size           scaling factor of the element
*/

/* description of the used element; the fixed size of the rectangle is
   scaled due to the IFS–process                                        */

ELEMENT recta  (pos: POSITION; angle: FLOAT;  size: FLOAT) :
                                                           COLOR;

BEGIN
    BOUNDARY              =  TRANSLATE ( pos,
                             ( ROTATE ( angle ,
                             ( SCALE (size,
                             RECTANGLE (0.8 0.6) )));

    CONTENTS.COLOR = ACNS (BLUEISH)
END;

/*  description of the resulting texture                             */

BEGIN
    CONTENTS = SCATTER ( RECURSIVE ( (ifs–code) , nsteps))
END
```

5.6 Conformity of Fractals Textures with the Texture Model

We have mentioned that it would be a better solution to specify equations at the texture synthesis language level for a universal construction algorithm integrated in HiLDTe.

Other problems arise from the requirements of the texture definition to textures. The textures should be specified as location–dependent functions, which are unlimited in space

In the following paragraphs, we present the different aspects of the term "texture" and sign up the way how conformity could be reached, if possible.

Unlimited definition space

As mentioned above, the texture functions should be defined for the unlimited space. But this is not reachable for all fractals. Imagine the Mandelbrot set: Here only special regions are of interest for fractal textures. The only way to have a texture at a fitting size is to scale the texture to the desired size.

Independence of resolution

Some fractal generation algorithms calculate the fractal model only for discrete points. Like for the midpoint subdivision, the calculation of the itensity value for a coordinate between two given points only can be gained by means of an interpolation function.

If the renderer needs such points for texture mapping, the easiest solution is to interpolate such points.

Attachability

For the rendering process, some other requirement are to be met. In order to map textures onto objects, it is important that the texture could be scaled to save the three–dimensional effect on such scenes. The borders of the texture must fit together. This is necessary, i.e., for mapping a texture onto a cylinder where it should not be possible to recognize the beginning or ending of the texture.

It is quite hard to meet this requirement for non–fractal textures. For fractal textures, it is just possible to apply special methods like linear interpolation between boundary fields of the attached textures.

Scaleability

Scaling fractal textures must be performed with care. If the fractal and not the representation of the fractal at a given resolution is scaled, the scaled texture could appear in a completely different manner. To avoid this, only the model of the fractal should be scaled.

Homogeneous appeareance

The homogeneity of the apparance of fractal textures varies with the type of fractal and the used algorithms for mapping it into the texture space.

The homogeneity of the appeareance could be measured with the texture gradient.

6 Conclusions

The paper presents the integration of fractal algorithms into a sophisticated texture specification and synthesis system. This texture editor uses a powerful formal texture specification language, HiLDTe. It is possible to integrate almost all known texture generation algorithms into HiLDTe, at least by means of user–defined processes. In spite of the recursivity and the limited definition space of most of the fractal algorithms – – properties which are contradictory to the given definition of the term texure – – the requirements for an integration of fractals into HiLDTe have been met.

However, an efficient and comfortable handling of fractals in texture synthesis demands for an intelligent support at the user interface level. It is planned to develop an object–oriented application surface with the help of artificial intelligence techniques.

Acknowledgements

The authors wish to express their thanks to Professor J. Encarnacao and to their colleagues and students, especially to W. Müller , J. Herder, V. Jung, and D. Lanio, and finally to S. Wurster for proof-reading.

References:

[Ba*etal*88] Barnsley, M. F., Jacquin, A., Malassenet, F., Reuter, L., Sloan, A. D.: Harnessing Chaos for Image Synthesis, Computer Graphics, Vol. 22, No. 4, pp. 131–140, 1988

[BaEl88] M. Barnsley, J. Elton: A New Class of Markov Processes for Image Encoding, Lournal of Applied Probality, No.20, pp. 14 –32, 1988

[BeJu83] Bergen, J.R., Julesz, B.: Rapid Discrimination of Visual Patterns, IEEE Transactions on Systems, Man and Cybernetics, Vol. SMC–13, No. 5, pp. 857–863, September/Oktober 1983

[DeHN85] Demko, S., Hodges, L., Naylor, B.: Construction of Fractal Objects with Iterated Function Systems, ACM Computer Graphics, Vol. 19, No. 3, pp. 271–278, 1985

[E*etal*90] Encarnacao, J.L, Dai, F., Englert, G.,Krömker, D., Sakas, G.: Der Textureditor; Ein System zur Generierung, Manipulation und Archivierung von Texturen, R&D–report GRIS 90–3, Technical University of Darmstadt, FB Comuter Science, FG Graphics Interactive Systems, April 1990 (in German)

[EnHS88] G. Englert, G. R. Hofmann, G. Sakas: Ein System zur Generierung und Archivierung von Texturen – Textureditor –, , in: Barth (Ed.): Visualisie–rungstechniken und Algorithmen, Informatik–Fachberichte , No. 182, pp. 155–173, 1988 (in German)

[EnSa89] G. Englert, G. Sakas: A Model for the Description and the Synthesis of Heterogeneous Textures, Proceedings of Eurographics, September 1989, pp. 245–256, 1989

[Falc85] Falconer, K.: The Geometry of Fractal Sets, Cambridge University Press, 1985

[Gard85] Gardner, G.Y.: Visual Simulation of Clouds, ACM Computer Graphics, Vol. 19, No. 3, pp. *297–303, 1985*

[HaBa84] Haruyama, S., Barsky, B.: Using Stochastic Modeling for Texture Generation, IEEE CG&A, pp. 7–19, 1984

[Herd90] Herder, J.: Konzepte und Implementierung einer Textur–Synthese–Sprache, Studienarbeit, Technical University of Darmstadt, FB Informatik, FG Graphisch–Interaktive Systeme, März 1990 (in German)

[Jule75] Julesz, B.: Experiments in the Visual Perception of Texture, Scientific American, Vol. 4, pp. 34–43, 1975

[Kauf88] Kaufman, A.: TSL – a Texture Synthesis Language, The Visual Computer, Vol. 4, pp. 148–158, September 1988

[KöBE90] Köhler, D., Baumann, P., Englert, G.: Das Texturarchiv als Beispiel für den Einsatz nichtkonventioneller Datenbanktechniken, Proc. Int .Workshop on Intergrated Intelligent Information Systems, Schloß Tucczno, Pila, Polen, pp. 156–175, September 1990 (in German)

[Krue88] Krüger, W.: Intensity Fluctuations and Natural Texturing, ACM Computer Graphics, Vol . 22, No.4, pp. 213–220, 1988

[LaKa87] Lauwerier, H., Kaandorp, J.: Fractals: Mathematics, Programming and Applications, Tutorial EUROGRAPHICS '87, Amsterdam, 1987

[Mand82] B.B. Mandelbrot: The Fractal Geometry of Nature, W.H. Freeman and Company, New York, USA, 1982

[MiRo89] M. Mischelitsch, O.E. Rössler: A New Feature in Hénon's Map, Computers & Graphics, Vol. 13, No. 2, pp. 263 – 265, 1989

[Noth85] Nothdurft, H.C.: Sensitivity for Structure Gradient in Texture Disrimination Tasks, Vision Research, Vol. 25, No. 12, pp. 1957–1968, 1985

[PaGo90] Papathomas, Th. V., Gorea, A.: The Role of Visual Attributes in Texture Perception, SPIE, Vol. 1249, Human Vision And Electronic Imaging: Models, Methods, And Applications, pp. 395 –403, 1990

[PeSa88] H.–O. Peitgen, D. Saupe: The Science of Fractal Images, Springer, New York, Berlin, Heidelberg, 1988

[Pent84] Pentland, A.P.: Fractal–Based Description of Natural Scenes, IEEE Transactions on Pattern Analysis and Machine Inteligence, Vol. PAMI–6, No. 6, pp. 661–674, November 1984

[Sche91] M. Schendel: Die fraktale Geometrie und ihre Anwendung zur Generierung von naturähnlichen Texturen, Strukturen und Objekten, diplom thesis, Technical University of Darmstadt, FB Informatik, FG Graphisch–Interaktive Systeme, 1991(in German)

[Smit84] Smith, A.R.: Plants, Fractals and Formal Languages, ACM Computer Graphics, Vol.18, No. 3, pp. 1–10, 1984

[Trei85] Treisman, A.: Preattentive Processing in Vision, Computer Vision, Graphics, and Image Processing, Vol. 31, pp. 156–177, 1985

[Upso86] Upson, C.: The Visual Simulation of Amorphous Phenomena, The Visual Computer, Vol.2, No.5, pp. 321–326, September 1986

[WMN85] Wyvill, B; Mcpheeters, C.; Novacek, M.: High Level Descriptions for 3D Stochastic languages, in: Magneat, Magneat–Thalman (Hrsgs.): Computer Generated Images, Proceedings of Graphics Interface '85, Springer Verlag Tokyo Berlin, 1985

Boundary Tracking of Complicated Surfaces with Applications to 3-D Julia Sets

C. Zahlten

Institut für Dynamische Systeme, Universität Bremen

Abstract

Cross sections of Julia sets in the quaternions are highly complicated 3-dimensional objects, which may serve as qualified test objects for surface construction and rendering algorithms. Boundary tracking methods generate the surfaces as lists of primitives such that rotation or repositioning requires only re-rendering. The main disadvantage of early boundary tracking approaches is the amount of storage they require. Ray-tracing methods, although adapted to the fractal setting, are time consuming since the objects have to be generated anew for each rendering. The *Chain of Cubes* algorithm used in this article is a boundary tracking method which uses a minimum of storage to generate a polygonal approximation of an iso-valued surface.

1 Introduction

The computer graphical interest in Julia sets is motivated by several properties of these objects: they are generated by simple rules; their shapes may show any degree of complexity; and illustrations of Julia sets and the Mandelbrot set are by now well known [MAN82, PEI86]. This makes Julia sets an ideal example to test the strength and weakness of algorithms which render complicated surfaces, although they are designed to operate in a more general setting.

In 1982 A. Norton presented an algorithm for the generation and display of 3-dimensional geometric fractals [NOR82, NOR89]. Two kinds of 3-D objects were shown. The first one was defined as a stack of 2-dimensional parameterized fractal curves, where each slice results from iterating a quadratic polynomial in the complex plane. The shape of the object shows the effect of varying the parameter. The second kind of fractals was given by extending the iterative functions to a higher dimensional space, the space of quaternions $\mathbb{H}$. Such quaternion Julia sets are the subject of this article. Since $\mathbb{H}$ is a 4-D space, 3-dimensional cross sections of the objects are visualized.

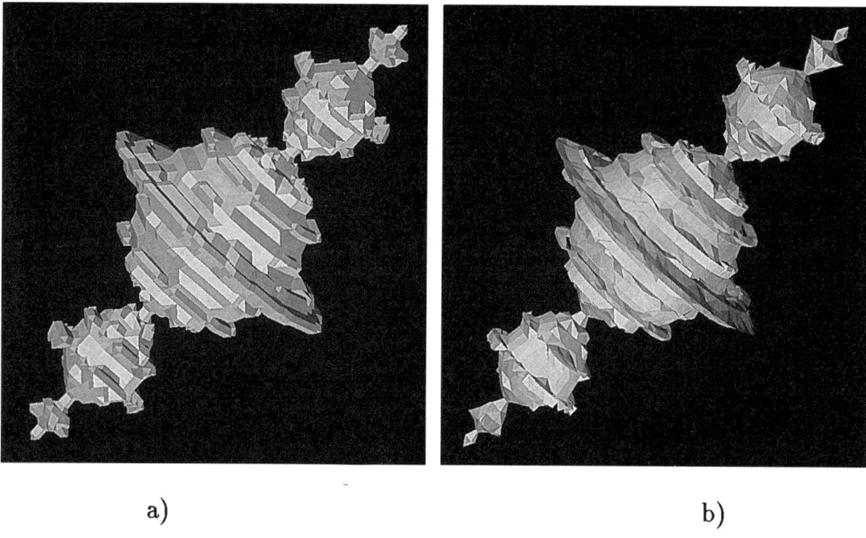

Figure 1: Julia set with real parameter $c = -1$ generated by cubes of size $\delta = 0.06$; a) interpolation, b) 4 bisection steps.

3 Generation and Rendering of the Surface

The *chain of cubes* has been described in [JÜR89] and [ZAH91] and we therefore only recall the basic outline of the algorithm. For approximating the surface, a tiling of space into cubes is used, such that the function is evaluated only on grid points. A cube is called transverse, if at least one of its vertices lies within the object and if at least one vertex lies on the outside. Starting with an initial transverse cube, the algorithm proceeds like a wave or like a rubber band which is pulled over the surface and traces one connected component. A complete list of all transverse cubes is written to an output file. All cubes within the "wave front" are held in a work list. In order to check whether a cube has already been determined, it is not necessary to look for it in a list of all previously found cubes. It is sufficient to search within the much smaller work list. The work is done, when the wave front either leaves the region of interest or collapses to nothing (i.e. the list is empty). The algorithm may be started with a set of initial cubes. It then tracks the boundary of all connected components with at least one initial cube on them.

From the cubes a consistently oriented polygonal approximation is generated. In a first approach the midpoints of the transverse edges of the cubes are connected by appropriate polygons to approximate the surface pieces within the

a) b)

Figure 2: Julia set with a) real parameter c = -1 and b) complex parameter $c = -1 + i \cdot 0.15$; stepsize $\delta = 0.002$.

cube. The triangles which build the polygons, then have only a small number of different directions. Therefore, only a few normals have to be calculated and real-time shading is possible by using fast look-up table shifts. An improved approximation of the surface is obtained by calculating the intersection of surface and cube more precisely. In this case standard polygon rendering is used. If a smooth surface is considered, an interpolation of the function values at the vertices is used to approximate the intersections of surface and edges. For the non-smooth surfaces of fractal objects, bisection is a better solution. The function is evaluated at the midpoint of the edge in question, and the transverse part of the edge is then chosen for the next bisection step. Although the evaluation of the iterative function is expensive, fractals require several bisection steps. In the current implementation it is possible to choose from 0 up to 8 bisection steps. Using bisection, the output file contains only one byte for each transverse edge (of a transverse cube) to hold the "address" of the approximated intersection points. Whereas for interpolation, it is necessary to hold for each transverse edge one floating point number between 0 and 1, but of course no extra space is needed if no approximation of the intersection point is chosen.

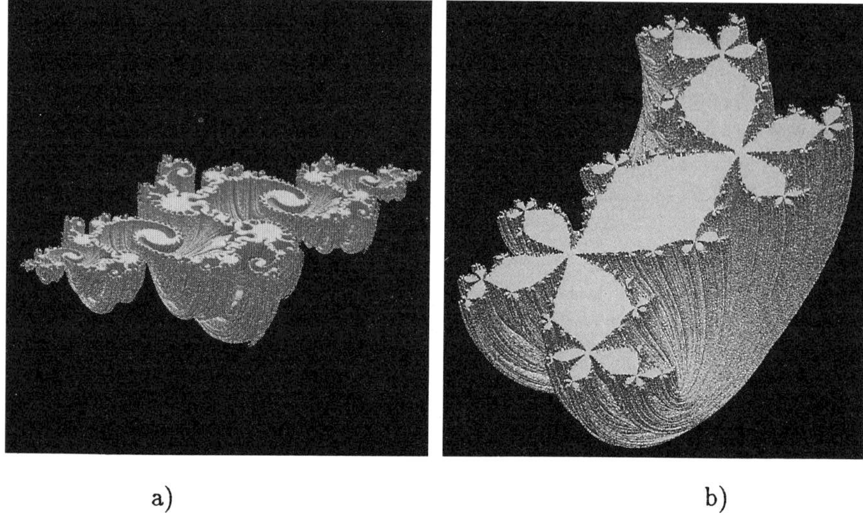

a) b)

Figure 3: Julia set with $c = 0.2809 + i \cdot 0.53$ a) generated by cubes $\delta = 0.005$, b) by ray-tracing

4 Examples

The illustrated Julia sets were generated and displayed on an IRIS 4D/25 graphics workstation with a single 20 MHz MIPS R3000 processor. If the polygonal approximation is done by simply using the midpoints of the cube edges, it takes about 60 minutes to generate objects which consist of about 8 million cubes. They are displayed within 9 minutes. For an improved approximation which uses 8 bisection steps, the generation time increases by a factor of about 4. Moreover, the display time increases since now the surface normals have to be calculated for each triangle.

Julia sets with real parameters $c \in \mathbb{R}$ possess a rotational symmetry with respect to the real axis. Figure 1 shows the Julia set according to $c = -1$ and illustrates the difference of using interpolation (1a) or 4 bisection steps (1b) to approximate the intersection of cubes and surface. In both cases 4000 cubes of size 0.06 were found by the *chain of cubes*. As described above, the 3-D object is calculated within the 3-D subspace $X \subset \mathbb{H}$.

If the parameter is slightly changed from $c = -1$ to $c = -1 + i \cdot 0.15$ the rotational symmetry is lost, as illustrated in figure 2. Both shapes are generated by 8 million cubes of size $\delta = 0.002$, and 8 bisection steps are used.

An example of a Julia set with an attractive 4-cycle is shown in figure 3a. As the pictures above, this image is generated by the *chain of cubes*, but this time no bisection steps are used. The object is cut along the complex plane such that the complex Julia set is visible. It is generated within 64 minutes, rendered within 14 minutes and real-time shading is possible. Figure 3b illustrates the same Julia set, generated with a modified version of the ray-tracing algorithm by

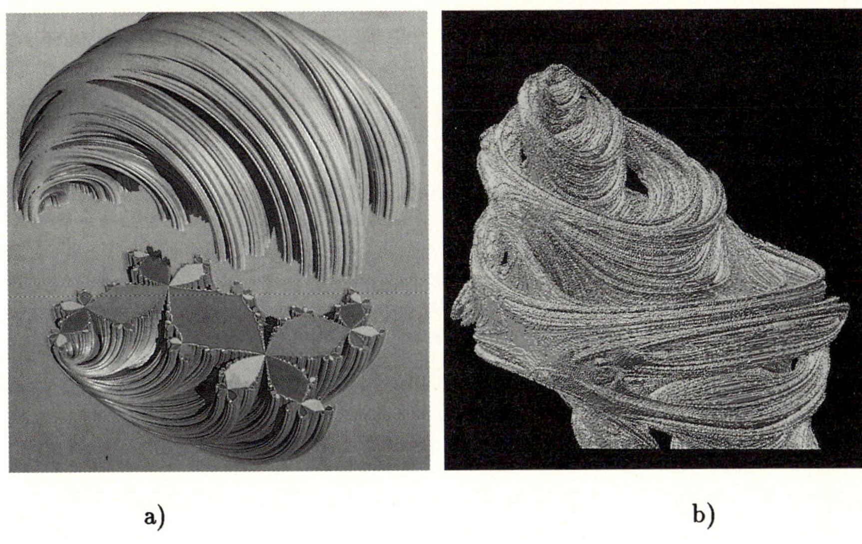

Figure 4: Julia set with a) $c = -0.12 + i \cdot 0.77$, $\delta = 0.005$, no bisection steps; b) $c = -0.77568377 + i \cdot 0.13646737$; $\delta = 0.005$; 4 bisection steps.

J. Hart et al. [LIC91]. The four colours of the cutting plane denote the basins of attraction of the four periodic points. Rendering took more than 2 hours.

Figure 4a) shows a rough approximation of $J_c(f)$ with $c = -0.12 + i \cdot 0.77$, using cubes of size $\delta = 0.005$. In Figure 4b the parameter is set to $c = -0.77568377 + i \cdot 0.13646737$ and the object is cut along the complex plane. Using cube size $\delta = 0.005$ and 4 bisection steps, it is generated within 64.4 minutes. About 2 million cubes intersect the surface. The rendering of this last example was done using the GRAPE visualization tool [RUM91] and took 12 minutes.

5 Conclusion

The *chain of cubes* algorithm gives the possibility of rendering complicated 3-D objects within reasonable time, even if they consist of millions of triangles. Nevertheless, the quality of the images is generally not comparable to those generated by the more time-consuming ray-tracing. For getting an idea of what a Julia set looks like, the *chain of cubes* is a suitable and fast solution. A future visualization tool for complex 3-D objects therefore may contain the *chain of cubes* for object generation and interactive selection of its positioning, and a (fractal) ray-tracer for high-quality visualization.

6 References

[ALL85] E.L. Allgower, P.H. Schmidt: An Algorithm for Piecewise Linear Approximation of an Implicit Defined Manifold, SIAM J. Numer. Anal. 22, 322-346, 1985

[ALL87] E.L. Allgower, S. Gnutzmann: An Algorithm for Piecewise Linear Approximation of Implicitly Defined Two-Dimensional Surfaces, SIAM J. Numer. Anal. 24, 452-469, 1987

[ALL90] E.L. Allgower, K. Georg: Introduction to Numerical Continuation Methods, Springer-Verlag, New York, 1990

[HAR89] J.C. Hart, D.J. Sandin, L.H. Kauffman: Ray Tracing Deterministic 3-D Fractals, Computer Graphics (Proc. SIGGRAPH) 23(3), 289-296, 1989

[JÜR89] H. Jürgens: Optimierte Oberflächenabtastung mit orientierten Kubusketten, in H. Jürgens, D. Saupe (Eds.): Visualisierung in Mathematik und Naturwissenschaften, 53-66, Springer-Verlag, Heidelberg, 1989

[LIC91] R. Lichtenberger: Visualisierung von Fraktalen mit Raytracing, Diplomarbeit, Universität Bremen, 1991

[LOR87] W.E. Lorensen, H.E. Cline: Marching Cubes: a High Resolution 3D Surface Construction Algorithm, Computer Graphics (Proc. SIGGRAPH) 21(4), 163-169, 1987

[MAN82] B.B. Mandelbrot: The Fractal Geometry of Nature, Freeman, San Francisco, 1982

[NOR82] A. Norton: Generation and Display of Geometric Fractals in 3-D, Computer Graphics (Proc. SIGGRAPH) 16(3), 61-66, 1982

[NOR89] A. Norton: Julia Sets in the Quaternions, Computers & Graphics 13(2), 267-278, 1989

[PEI86] H.-O. Peitgen, P. Richter: The Beauty of Fractals, Springer-Verlag, New York, 1986

[RUM91] M. Rumpf et al.: GRAPE GRAphics Programming Environment, Institut für Angewandte Mathematik der Universität Bonn, SFB 256, 1991

[WYV86] G. Wyvill, C. McPheeters, B. Wyvill: Data Structure for Soft Objects, Visual Computer 2(4), 227-234, 1986

[ZAH91] C. Zahlten: Piecewise Linear Approximation of Isovalued Surfaces, to appear in the Proceedings of the 2nd Eurographics Workshop on Visualization in Scientific Computing, Delft, The Netherlands, 22-24 April, 1991

3D-Rendering of Fractal Landscapes

H. Jürgens
Institut für Dynamische Systeme, Universität Bremen

Abstract

We discuss algorithms to render fractal landscapes and their application to the Mandelbrot set as well as Julia sets. For this application the integration of data generation and 3D-rendering provides a significant performance improvement.

1 Introduction

Fractals have become a fascinating part of computer graphics, although they cannot be considered to be part of the mainstream in computer graphics. The rendering of random fractal forgeries or the visualization of fractals of complex dynamical systems like the Mandelbrot set or Julia sets involve problems which have not been in the focus of computer graphical research. The typical basic primitives of computer graphics lines, circles, splines, polygons, etc. are not suitable for fractals, because one would need millions of theses for an acceptable resolution. Thus a special tailoring of algorithms for the rendering of fractals is required.

We begin our discussion with a simple algorithm for rendering fractal mountain forgeries (height fields). Variants of this algorithm were used not only to render the now famous images by Richard Voss (see [VOS85],[VOS88]) but also the first 3D-renderings of the potential of the Mandelbrot set. Common to these algorithms is a two step approach: first an equilateral grid of data (a height field) is generated, then this array is rendered. The optimization of the algorithm for the visualization of the Mandelbrot set (as well as for the 3D-rendering of Julia sets) is the main topic of this article. It involves the integration of both steps and tries to avoid the computation of unnecessary data (i.e., avoid the computation of the height field at those parts which become invisible in the course of the rendering process). This employs a predictor/corrector strategy which is based on an estimation of the distance to the Mandelbrot set (resp. Julia set).

The method of distance estimation already has been used by Fischer [FIS88] for fast algorithms for b/w-pictures of Mandelbrot- and Julia sets and also by Saupe

[SAU89] for his amazing pseudo-3D-images. We will discuss a variation of Fisher's algorithm which allow the fast computation of continuously shaded 2D-color images (where the color codes the distance to M) before we turn to the fast rendering of 3D-images (where the height represents the distance or the potential). The latter algorithms have been used for the production of [PJZ90] and in the latest version of [PJS90]. In this article we will restrict ourselves to the application to the Mandelbrot set. For the generation of images of Julia sets minor modifications of the algorithms are required. They have been implemented in [PJS90], but we skip the details.

2 Rendering of height fields

We assume that we have computed an n by m height field (i.e., for an equilateral grid of data points we have computed a height value at each data point which we enter into an n by m matrix of floating-point numbers). This could be obtained by an algorithm generating a random fractal or by evaluating the potential of the Mandelbrot set (see further below). Our approach to render such a height field is based on the idea of a "floating horizon" which is a well-known method of hidden-surface removal for vector graphics. Figure 1 illustrates the method: we scan the data (given as a grid of, say, 50 by 50 data points) row by row from front to back. For each slice we draw the projection of the data as the outline of a polygon or rather we draw those parts of the outline which supersede the data which has been drawn so far. The visibility (superseding or not) is tested by comparing the data with the floating horizon which is simply the polygon which just encloses the polygons drawn so far.

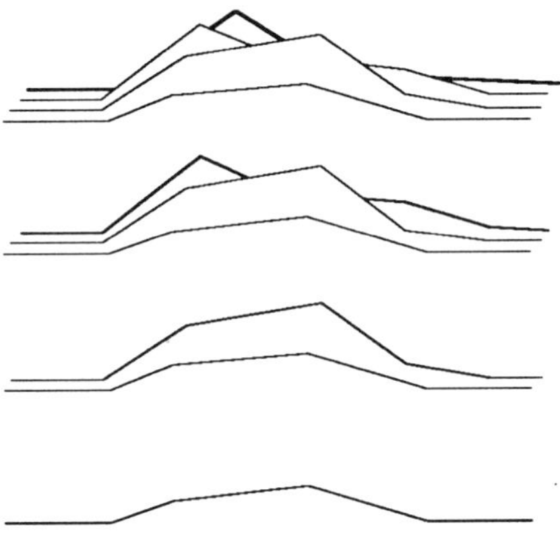

Figure 1: Floating horizon method using polygons

The algorithm *Height3D* implements this idea in a slightly different way, suitable for data arrays of higher magnitude (in the order of the size of a high-resolution computer screen, say, 1000 by 1000). Again we scan the data row by row from front to back. Now we assume, that we use a projection aligned to the data grid. In this way the problem of visibility can be solved for each column of the data grid independently (see figure 2). In the actual algorithm we use the vector *hor[]* (its size corresponds

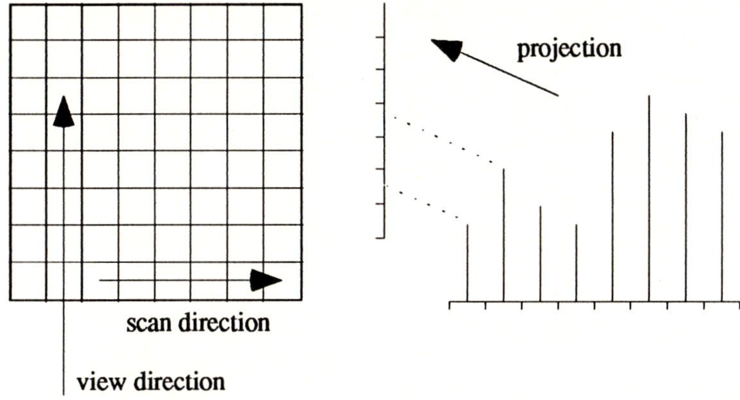

Figure 2: Projection parallel to the columns of the data grid simplifies the visibility problem: columns can be considered independently

to the width of the data array) to represent the height of the floating horizon for each column. Now let us consider the rendering of just one column of data.

First the front most data point is projected to the viewing plane. The projected height is used to initialize the value of the horizon. Furthermore we determine a normal vector at the data point (e.g., this can be an approximation based on the linear interpolation through two of the neighbour data points), compute the shading (resp. lighting) of the point based on the normal vector and the setting of light sources [1] and enter the result into the output array. When considering the next data point of the column, its projected height is compared with the horizon. If it does not supersede the horizon we simply continue with the next data point. Otherwise we have detected visibility and we compute the lighting at the data point. The value for this point and the last point are linearly interpolated and the result is entered into the output array (clipped to the entries $hor[ix]$ and $m2$). Finally the horizon is updated to its new value.

This algorithm can be easily extended to the generation of shadows. We only have to restrict to directional light sources whose direction is parallel to the data grid. Assume, for example, that we want to consider a light source positioned to the left of our data grid. If we restrict to the case that the light direction is parallel to the rows of the grid, the problem to check whether a data point is in shadow or not can be solved for the data points of each row independently. This can be done in the very same way as we solved the visibility problem: we introduce a floating (shadow) horizon for the light source. If the projected height of the data point (using the direction of the light source) does not supersede the horizon it is in shadow. Otherwise the light

[1] With LINTPOL we have implemented the (linear) interpolation of the coloring according to a rather simple but effective lighting model which is evaluated at the endpoints of the interpolation interval. The model uses three directional light sources whose vertical direction can be specified freely (horizontally they are aligned to the data grid), and one helmet-lamp (i.e., a light source which is located at the position of the viewer). The shading is computed from the scalar product of the normal vectors and the light vectors in the usual way (ambient light plus diffuse reflection).

```
ALGORITHM Height3D (Out,n, m, beta,light...)
Title         Compute one column of a 3D-rendering of M

Arguments     Out[n][m2]    the output array of colors for this column
              beta          vertical viewing direction
              light...      light parameters
Variables     ihor[n]       the horizon initialized to 0
Functions     FIELD(ix,iy) computes the value of a nx by ny height field at (ix,iy)
              NORM(Nx,Ny,ix,iy) computes the normal vector (Nx, Ny) at (ix,iy)
              PROJECT1(H,iy,beta) computes the projection onto viewing plane
              LINTPOL(Outp,iy,iynew,h, m2,Nx,Ny, Nxnew,Nynew) computes lighting and
                            interpolates values for output array (from value at Outp[iy] to the
                            new value at Outp[iynew]), output clipped at iy = h and iy = m2

BEGIN
   FOR ix = 1, n  {this loop initializes the interpolation}
      H = FIELD(ix,1)
      ip[ix] = PROJECT1(H,1,beta)
      IF (ip[ix] > ihor[ix]) THEN
         NORM(Nx[ix],Ny[ix],ix,iy)
         hor[ix] = ip[ix]
         LINTPOL(Out[ix],ip[ix],ip[ix],0,m2,Nx[ix],Ny[ix],Nx[ix],Ny[ix])
      END IF
   END FOR
   FOR iy = 2, m
      FOR ix = 1, n
         H = FIELD(ix,iy)
         ipnew = PROJECT1(H,iy,beta)
         IF (ipnew > ihor[ix]) THEN  {this is the visibility test}
            IF (ip[ix] ≤ ihor[ix]) NORM(Nx[ix],Ny[ix],ix,iy-1)
            NORM(Nxnew,Nynew,ix,iy)
            LINTPOL(Out[ix],ip[ix],ipnew,ihor[ix],m2,Nx[ix],Ny[ix],Nxnew,Nynew)
            hor[ix] = ipnew
            Nx[ix] = Nxnew
            Ny[ix] = Nynew
         END IF
         ip[ix] = ipnew
      END FOR
   END FOR
END
```

source contributes for the shading of the data point. The effect of this technique can be observed in figure 3 which shows a rendering of the potential of a part of the Mandelbrot set (for the potential function see further below).

Note that by scanning the data row by row the algorithm leads naturally to an output which is also organized row by row. This is convenient for most output devices. On the other hand, this implies that we have to work with a rather large output array. Scanning column by column is more suitable to solve the visibility problem. Organizing the algorithm in this way obviously reduces the memory requirements significantly (also, it becomes a little faster). But this requires either an output device which can draw column by column or the post-processing of the data (i.e., for devices which can only draw row by row transposing of the output data would be necessary).

Figure 3: Potential of Mandelbrot set rendered by an algorithm similar to *Height3D* (including shadow).

3 Distance estimator algorithms for the Mandelbrot set

The computation of the Mandelbrot set M (see section 5) by the ordinary pixel-scanning algorithm is widely known. The pixels of an image are scanned row by row and column by column. Each pixel is colored depending on the result of an iteration count, the computed potential or the distance to the Mandelbrot set M (in fact, the method can also be used to generate a height field which then is rendered by our algorithm *Height3D*). But this approach is rather (cpu-)time consuming. In the book "The Science of Fractal Images" Y. Fisher [FIS88] proposed a fast algorithm for b/w-pictures of M (and also for Julia sets J_c). It is based on the estimation and bounding of the distance of a given point to the set. Very roughly this algorithm is 2 to 5 times faster than scanning algorithms. In "The Game of Fractal Images" [PJS90] Fisher's algorithm was extended to allow the production of color images by introducing a simple coloring scheme for the disks which are generated by the algorithm. An interesting modification of this algorithm by D. Saupe [SAU89] created amazing pseudo-3D-images, showing M embedded in thousands of small spheres, which correspond to the disks generated by the original algorithm.

We discuss a variation of Fisher's algorithm which allows the fast computation of continuously shaded 2D-color images (where the color codes the distance to M) and then extend the algorithm to the fast rendering of 3D-images (where the height represents the distance or the potential). These algorithms scan the data column by column. In the following we therefore only discuss the computation of one such column. Algorithm *DistShade* is the sketch of a program describing the generation of a color image, which is continuously shaded with respect to the distance to M. It is interesting to note that this algorithm is much simpler to program than the original algorithm by Fisher. Yet both algorithms exhibit approximately the same saving of computer time. In fact, it very much depends on the specific example image and the hardware to decide which one is faster.

```
ALGORITHM DistShade (Out,x, m, ymin, ymax)
Title      Compute one column of a continuously shaded image of M (using distances)

Arguments  Out          the output array of colors for this column
           x            x-value of column
           m            image resolution in y-direction
           ymin, ymax   low and high y-value of image
Functions  MDIST(inM,x,y) returns the distance to the boundary of M measured in
                        pixels and sets inM (if the point is in M)
           CINTPOL(Out, iy, iynew, D, Dnew) sets the array out[iy] to out[iynew]
                        to the linearly interpolated color coding of D to Dnew
           BLACK(Out,iy,iynew) sets the array to black
           Min(), Max() the minimum/maximum of two numbers

BEGIN
    iy =0
    D = MDIST(inM,x,ymin)
    WHILE (iy < m)
        iynew = iy + Max(1,Min(20,D))
        y = iynew*(ymax-ymin)/(m-1)
        Dnew = MDIST(inM,x,y)
        IF (inM) THEN
            BLACK(Out,iy,iynew)
        ELSE
            CINTPOL(Out, iy, iynew, D, Dnew)
        ENDIF
        iy = iynew
        D = Dnew
    END WHILE
END
```

The algorithm avoids the computation of the distance to M (see further below) for most pixels of the column. It performs these calculations only for certain node points. From one such point (represented by a pixel) it tries to compute (using the distance estimate to M) another point such that the connection of both points does not intersect the boundary of the Mandelbrot set (otherwise they will be represented by two pixels which lie next to each other). The values for all pixels between these two points are obtained by linear interpolation. The error which results from the interpolation (vs. the precise distance values) can be bound by controlling the maximal step size. The worst case will occur when the algorithm passes dendrites (we sketch the situation in figure 4). Here our interpolated values can be off by the factor $1-\sqrt{3}/2 \approx 10\%$. Setting the maximal steps ize to 20 pixels (as shown in algorithm DistShade) limits the maximal error (the difference between distance estimation by interpolation vs. explicit distance computation) to approx. 2 pixels. In practice this eliminates all visually detectable artefacts which result from interpolation errors. On the other hand bounding the maximal step size to this value does not effect noticeably the performance of the algorithm.

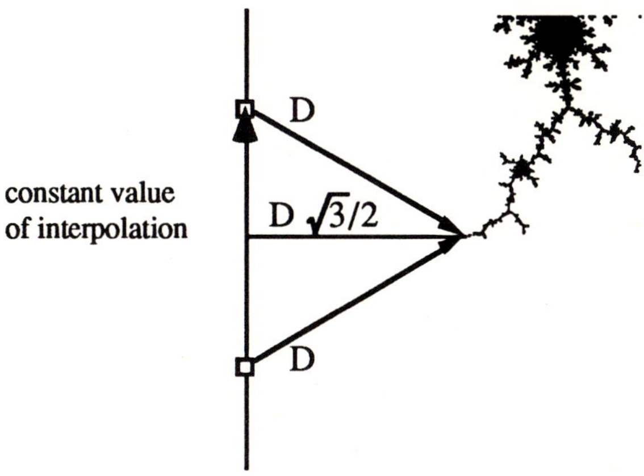

Figure 4: Interpolation error when the algorithm passes dendrites

4 Fast rendering of 3D-images

The integration of a step size control as in *DistShade* and the rendering algorithm of *Height3D* leads to our algorithm Dist3D. It is the outline of a program which allows to compute 3D-renderings of the Mandelbrot set. The algorithm scans the data by going step by step from near to far to a given viewing direction. Again the visibility problem is solved by comparing to a floating horizon (i.e., we do not draw what is hidden behind the current horizon but we draw what is visible and the visible part determines the new horizon.

The function MDISTN does not only determine the distance to M, but it also computes the normal vector at the given position (which can be done explicitly). We will discuss these computations in more detail (see below). The function HEIGHT defines a mapping of the distance D to a certain height (e.g. $1 - D$). It also adjusts the normal vector according to this transformation. The function PROJECTION gives a parallel projection (onto the pixels of the image) and LINTPOL2 again provides the lighting computation and linear interpolation to compute the output as in *Height3D*. [2]

There are several problems which we do not address in the algorithmic description. Especially the visibility problem needs some care, because we do not want to increase the current point (x, y) in small steps but in jumps as big as possible (i.e., the amount given by MDISTN). Figure 5 shows two typical problems. Problem A: We jump from y_0 to y_1, missing a local maximum of the projected height. Thus we miss the correct horizon h at a. Problem B: We jump from y_1 (invisible) to y_2 (visible). The interpolation of the light model would be very poor in this situation, because the normal in y_2 is not appropriate. We would need the normal in b. Our complete

[2] The implemented lighting model uses two light sources, one directional light the direction of which can be specified freely, and one helmet-lamp (i.e., a light source which is located at the position of the viewer).

```
ALGORITHM Dist3D  (Out,x0,y0,radius, m, alfa,beta,light...)
Title         Compute one column of a 3D-rendering of M

Arguments     Out[m]        the output array of colors for this column
              x0,y0         base point of column
              radius        scaling factor for coordinate range
              alfa, beta    horizontal/vertical viewing direction
              light...      light parameters
Functions     MDISTN(Nx,Ny,inM,x,y) returns the distance to the boundary of M measured in
                            pixelsize/sin(beta), sets inM and Nx,Ny of the normal vector
              HEIGHT(Nx,Ny,D) computes the height from D and adjusts the
                            normal vector (Nx, Ny)
              PROJECT(H,x,y,beta) computes the projection onto viewing plane
                            (pixels of column)
              LINTPOL2(Outp,iy,iynew,h,m,Nx,Ny,Nxnew,Nynew) computes lighting and
                            interpolates values for output array (from Outp[iy] to the new value
                            at Outp[iynew]), values are clipped at iy = h and iy =m
              BLACK(Out,iy,iynew,h,m) sets the array to black

BEGIN
    x = x0; y = y0
    xd = cos(alfa) * radius / m / sin(beta)
    yd = sin(alfa) * radius / m / sin(beta)
    D = MDISTN(Nx,Ny,inM,x,y)
    H = HEIGHT(Nx, Ny,D)
    iy= PROJECTION(H,x,y,beta)
    h = iy
    WHILE (iy < m)
        IF (D < 1) THEN
            xnew = x + xd; ynew = y + yd
        ELSE
            xnew = x + D * xd; ynew = y + D * yd
        ENDIF
        Dnew = MDISTN(Nxnew,Nynew,inM,xnew,ynew)
        H = HEIGHT(Nxnew, Nynew,Dnew)
        iynew = PROJECTION(H,xnew,ynew,beta)
        IF (iynew > h) THEN {this is the visiblility test}
            IF (inM OR Dnew < 0.5) THEN
                BLACK(Out,iy,iynew,h,m)
            ELSE
                LINTPOL(Out,iy,iynew,h,m,Nx,Ny,Nxnew,Nynew)
            ENDIF
            h = iynew
        ENDIF
        iy = iynew
        D = Dnew; Nx = Nxnew; Ny = Nynew
    END WHILE
END
```

implementation takes care of both problems. Point a is computed by an iteration scheme (using linear interpolations) which solves the equation $\langle N(x,y), L_H \rangle = 0$, where $N(x,y)$ is the normal vector in (x,y) and L_H is the light direction of the helmet-lamp! Point b is computed by an iteration which solves $P(x,y,H) - h = 0$ where $P(x,y,H)$ is the projection of a point (H is the height at (x,y)) which lies in the viewing direction (in fact, it is the first visible point, which is not hidden by

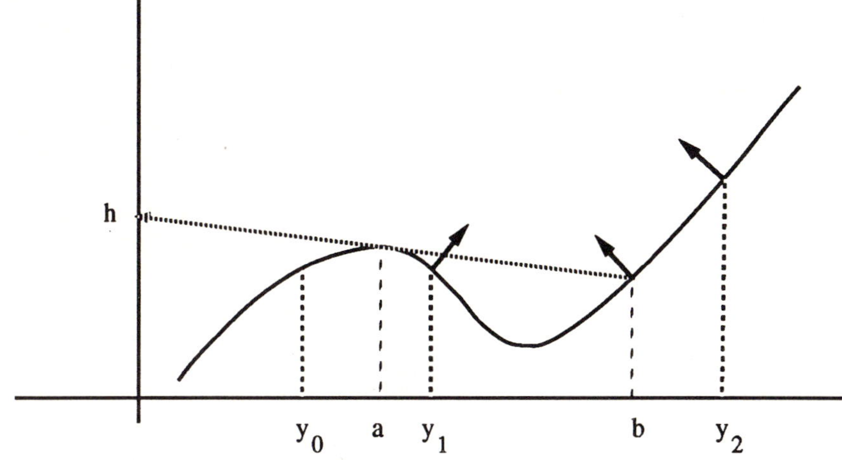

Figure 5: Interpolation nodes and floating horizon

Figure 6: Potential of Mandelbrot set rendered by an algorithm similar to *Dist3D* at ultra high resolution (12M pixels).

the horizon h). The actual algorithm employs also a step size control similar to the one discussed for the algorithm DistShade. But in this case we bound the step size only when we compute visible parts. This is essential for the performance of the algorithm. Otherwise in some situations we might waste enormous amounts of time in computing very precisely parts which, however, are hidden behind the horizon. We skip these technical details.

5 Distance estimation and normal vectors

The functions MDIST and MDISTN lie at the heart of our algorithms. They provide the estimate of the distance to M as well as the required normal vectors (MDISTN). Let us consider the iteration

$$z_k = z_{k-1}^2 + c, z_1 = c,$$

where z and c are complex numbers. The Mandelbrot set can be defined as

$$M = \left\{ z_1 | \lim_{k \to \infty} z_k \neq \infty \right\} .$$

The potential of the Mandelbrot set can be approximated by

$$G(c) = \frac{1}{2^n} \ln |z_n(c)|$$

if $z_n > 1/\epsilon$ (for a sufficiently small value of ϵ, e.g. 0.01). We set $G(c) = 0$ if $z_k < 1/\epsilon$ for all $k < N$ (where N is a sufficiently large number, e.g. 100). The functions MDIST and MDISTN estimate the distance by

$$D = \frac{G(c)}{2|G'(c)|} = \frac{|z_n| \ln |z_n|}{2|z_n'|}$$

(see also algorithm MSetDist in [PEI88]). For the following we define $z := z_n$ and write $c = x+yi$, $z = u+vi$. Furthermore let $(z, z')_x := u\frac{\partial u}{\partial x} + v\frac{\partial v}{\partial x} = \mathrm{re}\,z\ \mathrm{re}\,z' + \mathrm{im}\,z\ \mathrm{im}\,z'$ and $(z, z')_y := u\frac{\partial u}{\partial y} + v\frac{\partial v}{\partial y} = -\mathrm{re}\,z\ \mathrm{im}\,z' + \mathrm{im}\,z\ \mathrm{re}\,z'$. Then the normal vectors of the potential surface approximated by $G(c)$ can be obtained from

$$\frac{\partial G}{\partial x} = \frac{(z, z')_x}{2^n |z|^2}$$

and

$$\frac{\partial G}{\partial y} = \frac{(z, z')_y}{2^n |z|^2}.$$

For the computation of normal vectors of the surface representing the distance D we need several second derivatives. We have

$$\frac{\partial D}{\partial x} = (1 + \ln |z|) \frac{(z, z')_x}{2|z||z'|} - |z| \ln |z| \frac{(z', z'')_x}{2|z'|^3}$$

and

$$\frac{\partial D}{\partial y} = (1 + \ln |z|) \frac{(z, z')_y}{2|z||z'|} - |z| \ln |z| \frac{(z', z'')_y}{2|z'|^3}$$

with $(z', z'')_x = \mathrm{re}\,z'\ \mathrm{re}\,z'' + \mathrm{im}\,z'\ \mathrm{im}\,z''$ and $(z', z'')_y = -\mathrm{re}\,z'\ \mathrm{im}\,z'' + \mathrm{im}\,z'\ \mathrm{re}\,z''$. Moreover $z' := z_n'$ and $z'' := z_n''$ can be computed from the iterations

$$z_k' = 2z_{k-1}z_{k-1}' + 1, \ z_1' = 1$$

and

$$z''_k = 2(z'^2_{k-1} + z_{k-1}z''_{k-1}), \ z''_1 = 0.$$

Finally we can get an idea of the required computing time for these calculation by counting the floating point operations (FLOP). The simple iteration $z_k = z^2_{k-1} + c$ requires 7 FLOP per step. The computation of z'_k requires 9 additional FLOP and the computation of z''_k another 24 FLOP. Therefore the computation of G plus normal vectors (for the rendering of a 3D - potential surface) is nearly as fast as the continuous color shading algorithm (which does not need the normals). The 3D-rendering of a 3D-surface (which represents the distance D) is a little slower because we need approximately twice as many FLOP per iteration step. This, in fact, is a worst case estimate because for values c which lie in the Mandelbrot set we do not need the computation of z''_k. But in any case we have found that with these algorithms the computation and rendering of 3D-views of the Mandelbrot set is faster than the original scanning algorithm for 2D-images.

References

[FIS88] Y. Fisher, *Exploring the Mandelbrot Set*, in: Peitgen/Saupe, The Science of Fractal Images, Springer-Verlag, New-York, 1988

[PEI88] H.O. Peitgen, *Fantastic Deterministic Fractals*; in: Peitgen/Saupe, The Science of Fractal Images, Springer-Verlag, New-York, 1988

[PJS90] H.O.Peitgen/H. Jürgens/D.Saupe, *The Beauty of Fractals Lab* (The Game of Fractal Images Version 2.0), Macintosh Implementation by M.Parmet and T.Eberhard, Springer-Verlag, New-York, 1990

[PJZ90] H.O.Peitgen/H.Jürgens/D.Saupe/C.Zahlten, *Fractals: An Animated Discussion*, Video, 62 min, W.H. Freeman/Scientific American - Video, 1990

[SAU89] D. Saupe, *Turbo Mandelbrot Sets*, in: Fractal Report, (Reeves Telecommunications Laboratories, Cornwall, 1989), also in: Amygdala (a Newsletter of fractals and the Mandelbrot set), San Cristobal, New Mexico, 1989

[VOS85] R.F. Voss, *Random Fractal Forgeries*, in: Fundamental Algorithms for Computer Graphics, R.A. Earnshaw (ed.), Springer-Verlag, Berlin, 1985

[VOS88] R.F. Voss, *Fractals in Nature: From Characterization to Simulation*, in: Peitgen/Saupe, The Science of Fractal Images, Springer-Verlag, New-York, 1988

Fractal Interpolation of Random Fields of Fractional Brownian Motion

W. Rümelin

STN SYSTEMTECHNIK NORD,
Bremen, Federal Republic of Germany

Abstract

For the fractal interpolation of random fields of fractional Brownian motion (fBm), a scheme is proposed which allows the numerical generation of an arbitrary number of the values of fBm in one step on the basis of an arbitrary number of known values. The examples treat interpolation of random fields with regularly and irregularly spaced known values.

1 Introduction

Fractional Brownian motion (fBm) can be used for a mathematical description of many physical phenomena of a more or less random nature (e.g. modelling of mountains, clouds etc., see [MAN82], [VOS88]).

The algorithm presented here generates samples of random fields of fBm. It can be used to simulate landscapes on the computer where the gross shape is predetermined by a set of data points with given altitude and the roughness of the landscape is characterized by its (given or estimated) fractal dimension. The generated random field will pass precisely through the given data points and will have the prescribed fractal dimension. We call this a fractal interpolation of fBm. By successive subdivision of the underlying grid any wanted resolution can be achieved.

In a former paper [RÜM90] we have shown how to simulate one single new value of fBm at a time, while we now treat the more general problem of simulating arbitrarily many values all at once. This extension will facilitate the computation of fractal interpolations in the sense above. Related algorithms for special cases are treated in [FOU82], [FEL85], [LEW87].

2 Fractal Interpolation of fBm

2.1 Derivation of the scheme

Let a set of data points $\mathbf{P}_j = (\mathbf{t}_j, x_j) \in \mathbf{R}^3$, $j = 1, 2, \ldots n$ be given, where $\mathbf{t}_j \in \mathbf{R}^2$ resp. $x_j \in \mathbf{R}$ are called the coordinates resp. the value of the data point $\mathbf{P}_j$ ($\mathbf{R}$ denotes the space of real numbers). We want to generate realizations of a random field of fractional Brownian motion denoted by $x(\mathbf{t}) \in \mathbf{R}$, which interpolate the data. Sometimes the data are denoted as "old" values. We will generate the "new" values $y_i = x(\mathbf{s}_i)$ at the coordinates $\mathbf{s}_i$, $i = 1, 2, \ldots m$ in one step in such a way that the statistical properties of the assumed underlying fBm with given fractal dimension D are preserved (For more motivation and a definition of fBm see [RUM90]).

For this purpose we consider the scheme

$$(1) \qquad \mathbf{y} = \mathbf{B} \cdot \mathbf{x} + \mathbf{S} \cdot \mathbf{z}$$

where $\mathbf{y} = (y_1, y_2, \ldots y_m)^T$, $\mathbf{x} = (x_1, x_2, \ldots x_n)^T$, $\mathbf{B} = (b_{ij})$, $i = 1,2 \ldots m$, $j = 1,2, \ldots n$, $\mathbf{S} = (s_{ik})$, $i,k = 1,2, \ldots m$, $\mathbf{z} = (z_1, z_2, \ldots z_m)^T$ with components z_k being independent Gaussian random variables with mean zero and variance 1, i.e. $E\{z_k^2\} = 1$. While matrix $\mathbf{B}$ controls the influence of the history $\mathbf{x}$ of fBm, matrix $\mathbf{S}$ governs the rate of additional randomness necessary to produce the new values $\mathbf{y}$. The matrices $\mathbf{B}$ and $\mathbf{S}$ must be determined.

Since the underlying random field $x(\mathbf{t})$ is nonstationary but has stationary increments, we reformulate (1) in terms of increments of $x(\mathbf{t})$ with respect to an arbitrary but fixed reference value, say $x_n = x(\mathbf{t}_n)$, and consider

$$(2) \qquad y_i - x_n = \sum_{j=1}^{n-1} b_{ij} \cdot (x_j - x_n) + \sum_{k=1}^{m} s_{ik} \cdot z_k, \quad i = 1,2,\ldots m$$

which is equivalent to

$$y_i = x_n - \sum_{j=1}^{n-1} b_{ij} \cdot x_n + \sum_{j=1}^{n-1} b_{ij} \cdot x_j + \sum_{k=1}^{m} s_{ik} \cdot z_k, \quad i = 1,2,\ldots m$$

Thus

$$(3) \qquad b_{in} = 1 - \sum_{j=1}^{n-1} b_{ij}, \quad i = 1,2,\ldots m$$

i.e. the last column of $\mathbf{B}$ can be calculated given the other columns of $\mathbf{B}$.

Denoting $Dy_i = y_i - x_n$, $i = 1,2 \ldots m$ and $Dx_j = x_j - x_n$, $j = 1, 2, \ldots n-1$ we

(iii) Determine b_{in} using the b_{ij} just found,

$$(3) \qquad b_{in} = 1 - \sum_{j=1}^{n-1} b_{ij}\,, \quad i=1,2,\ldots m$$

(iv) Compute S from (8) by Cholesky decomposition,

$$(8) \qquad \mathbf{S}\cdot\mathbf{S}^T = \mathbf{C}_{yy} - \mathbf{B}'\cdot\mathbf{C}_{yx}{}^T.$$

(v) Generate m independent realizations of a Gaussian random variable with zero mean and variance 1 yielding a realization of the random vector **z**.

(vi) Compute the new values y according to

$$(1) \qquad \mathbf{y} = \mathbf{B}\cdot\mathbf{x} + \mathbf{S}\cdot\mathbf{z}$$

Notice that the scheme above (with modified $\mathbf{C}_{xx}$, $\mathbf{C}_{yx}$, $\mathbf{C}_{yy}$) can be used as well for the simulation of other types of Gaussian processes with stationary increments as long as the corresponding covariance matrix $\mathbf{C}_{xx}$ is positive definite.

Introducing a meshsize (or unit distance) h it is clear from (9) that $\mathbf{C}_{xx}$, $\mathbf{C}_{yx}$ and $\mathbf{C}_{yy}$ are proportional to h^{2H}. So **B** does not depend on h at all. On the other hand $\mathbf{S}\cdot\mathbf{S}^T$ is proportional to $\mathbf{C}_{yy}$, therefore **S** proportional to h^H. Thus, halving the meshsize h requires only a multiplication of **S** by a factor 0.5^H, $\mathbf{S}^{(h/2)} = 0.5^H\cdot\mathbf{S}^{(h)}$.

3 Application to the generation of random fields

The simulation of random fields without any prescribed values was already shown in [RÜM90] where one point is generated at a time. We restrict ourselves to the cases where some values of the random field are already known in advance and one wishes to interpolate through these values.

3.1 Regularly spaced data points: Subdivision

First, we treat the case where the data points are situated on a regularly spaced grid with fixed meshsize. We want to produce values of the random field on a grid with meshsize halved denoting this new meshsize by h.

Since the correlation between points decreases with increasing distance between points and on the other hand the computational effort grows with the number of data and new values the best method is to apply the algorithm locally on a neighborhood of the known points and to generate only three new points in one step. For the sake of simplicity we choose a square neighborhood area which is shaded (see Figure 1). The coordinates of the data points are marked with black and white dots. Data points inside the neighborhood are labelled with numbers from 1 up to 33. The new points to be generated lie in the center of the neighborhood, they are printed with a + sign and indexed with bold **1**, **2** and **3**. The white dots correspond to points generated before.

In order to fill up a mesh with interpolated points one must establish the matrices **B** and **S** (generically) only once, which can be done in advance. These matrices allow the computation of three new values corresponding to the neighborhood chosen. Shifting the neighborhood systematically over the plane in such a way that the configuration within the neighborhood remains the same one can apply the same **B** and **S** producing interpolated values. For the generation of points near the border where the neighborhood contains less data points **B** and **S** must be computed separately (cf. the discussion of subdivision in the one-dimensional case in [RUM90]).

The coefficient matrices **B** and **S** for the square neighborhood of Figure 1 with an exponent H = 0.7 (i.e. fractal dimension 2.3) are shown in Figure 2. The j-th row of $\mathbf{B}^T$ corresponds to data point j, while column i of $\mathbf{B}^T$ corresponds to new point number i. Looking at the values of the elements of **B** one can see how the weight of the points decreases with increasing distance from the corresponding new point. For a time-critical application

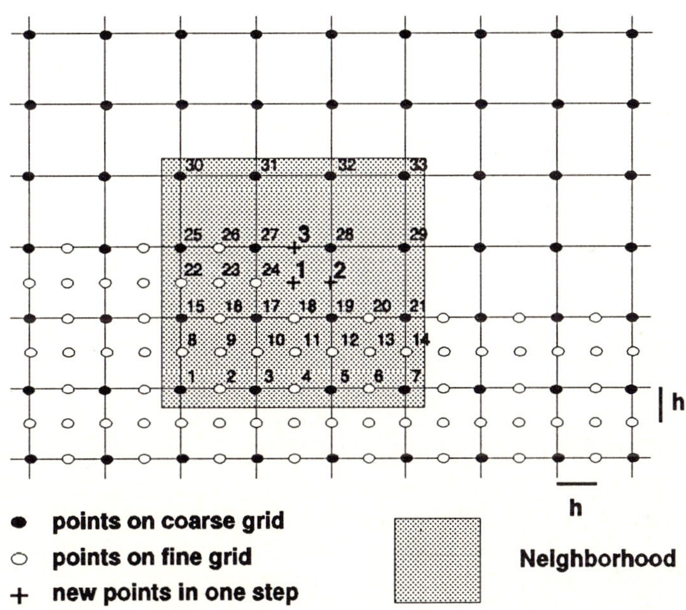

Figure 1: Neighborhood for the generation of three new points in one step

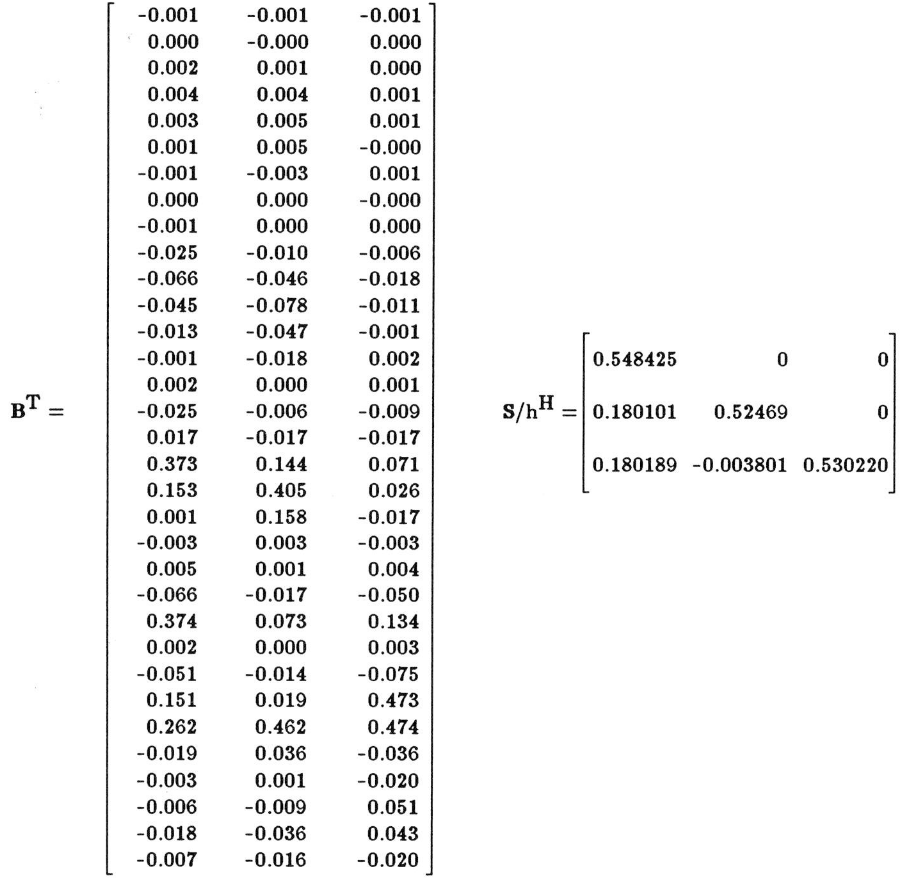

$$B^T = \begin{bmatrix} -0.001 & -0.001 & -0.001 \\ 0.000 & -0.000 & 0.000 \\ 0.002 & 0.001 & 0.000 \\ 0.004 & 0.004 & 0.001 \\ 0.003 & 0.005 & 0.001 \\ 0.001 & 0.005 & -0.000 \\ -0.001 & -0.003 & 0.001 \\ 0.000 & 0.000 & -0.000 \\ -0.001 & 0.000 & 0.000 \\ -0.025 & -0.010 & -0.006 \\ -0.066 & -0.046 & -0.018 \\ -0.045 & -0.078 & -0.011 \\ -0.013 & -0.047 & -0.001 \\ -0.001 & -0.018 & 0.002 \\ 0.002 & 0.000 & 0.001 \\ -0.025 & -0.006 & -0.009 \\ 0.017 & -0.017 & -0.017 \\ 0.373 & 0.144 & 0.071 \\ 0.153 & 0.405 & 0.026 \\ 0.001 & 0.158 & -0.017 \\ -0.003 & 0.003 & -0.003 \\ 0.005 & 0.001 & 0.004 \\ -0.066 & -0.017 & -0.050 \\ 0.374 & 0.073 & 0.134 \\ 0.002 & 0.000 & 0.003 \\ -0.051 & -0.014 & -0.075 \\ 0.151 & 0.019 & 0.473 \\ 0.262 & 0.462 & 0.474 \\ -0.019 & 0.036 & -0.036 \\ -0.003 & 0.001 & -0.020 \\ -0.006 & -0.009 & 0.051 \\ -0.018 & -0.036 & 0.043 \\ -0.007 & -0.016 & -0.020 \end{bmatrix} \qquad S/h^H = \begin{bmatrix} 0.548425 & 0 & 0 \\ 0.180101 & 0.52469 & 0 \\ 0.180189 & -0.003801 & 0.530220 \end{bmatrix}$$

Figure 2: Matrices **B** and **S** for the neighborhood of figure 1,
Exponent H = 0.7, Fractal Dimension 2.3

one could easily reduce the neighborhood by the points which have only a minor influence on the new values (e.g. the points with number 1 - 9, 15, 21, 22, 25). To be precise, one should compute **B** and **S** afresh with this altered neighborhood, but there will be only minor changes compared to the corresponding values of Figure 2.

Having filled up the fine grid with new points one could easily start again with a further subdivision of this grid using the same scheme and only adjusting the entries of **S** by a factor 0.5^H.

3.2 Irregularly spaced data points

If the data points are irregularly spaced it is not possible to apply such a shift-invariant set of matrices **B** and **S** as above. Since the configuration of the data points within the neighborhood of each new point will be different, it will be generally necessary to compute the matrices **B** and **S** anew for each neighborhood. If, in addition, the data points are sparsely distributed within the region considered, a practical way to achieve a true interpolation with a resonable computational effort is to generate at once all the new points on a regularly spaced grid (with a meshsize as small as computationally possible). This grid can be then refined with the method of subdivision from above (irrespective of the sparse old data points).

4 Example: Matterhorn

In order to demonstrate the power of our algorithm for fractal interpolation we have chosen the wellknown mountain "*Matterhorn*" located in the Swiss Alps.

We laid a grid of 250 m (in reality) meshsize over a map (scale 1:25000) of this mountain and estimated (by thumb) the values of altitude on the grid over a region of an original size of 3 by 3 km yielding 13 x 13 points. Applying our subdivision algorithm from section 3.1 we arrived at 385 x 385 points spaced with a meshsize of 7.8125 m (after a five-fold subdivision).

These 385 x 385 points are the geometric basis of our 3-dimensional interpolated picture in Figure 4. As a surface model we used the simple idea that in such a high alpine region there will be only two surface materials visible, namely snow or ice ("white color") and rock ("grey color") depending on the slope of the surface. We decided to present rock for a slope greater than 50 degrees and snow otherwise. In Figure 3 the input data (13 x 13 points) can be seen with the above surface model, while in Figure 4 the 385 x 385 points produced by our fractal interpolation (with fractal dimension 2.3) and the same surface model are shown. Both scenes are illuminated by a light source from the graphics library of a Silicon Graphics IRIS 4D. The viewing point is 4000 m above sea level (the top of the Matterhorn is 4478 m high), the viewing direction is from north-east. We display that part of the mountain which is above 2500 meters.

The overall performance looks rather realistic but the characteristics of the Matterhorn - the roof-shaped top and the steep eastern front side - are not at all modelled quite well. This should not astonish us since there is only one original data point situated in this particularly sensitive region.

In a second approach we therefore took as data points those points which are definitely marked as measure points in our map. There are only 47 such points in the 3 x 3 km region considered which are preferably located at ridges and local peaks. These irregularly and sparsely distributed points are augmented by another 7 points taken near the top of the mountain in order to emphasize the characteristics mentioned above.

On the basis of these 54 data points in one step we first calculated the interpolated values with the procedure from section 3.2 on a grid with meshsize 60 m yielding 51 x 51 values. Now, neglecting the old 54 points we feeded these 51 x 51 points into the subdivision algorithm from section 3.1 ending up with 401 x 401 points and a grid with meshsize 7.5 m (after 3 subdivisions). The resulting image is shown in Figure 5. Apparently, the top geometry of the Matterhorn is more realistic. Of course, due to the very few original data points the deviations from the real shape are greater in other parts of the mountain, but these defects are not so obvious.

5 Appendix

We prove that $S \cdot S^T$ given by equ. (8) is positive definite.

Since $S \cdot S^T$ is nonnegative definite by definition it is enough to show that $S \cdot S^T$ is nonsingular or has maximal rank m. If $S \cdot S^T$ has maximal rank, so has S which we are going to prove. From [RÜM90] we know that the covariance matrix of fBm is positive definite, thus the covariance of the vector formed by x and y, (x y) is positive, too.

On the other hand, positive definiteness of the covariance matrix is equivalent to the linear independence of the elements x_j and y_i (see [LAM77], p. 25), which means

$$\text{(A1)} \qquad \sum_{j=1}^{n} \alpha_j \cdot x_j + \sum_{i=1}^{m} \beta_i \cdot y_i = 0 \qquad \text{(with probability 1)}$$

if and only if all $\alpha_j = \beta_i = 0$.

Now, if we insert y from (1) into (A1) we get

$$\sum_{j=1}^{n} \alpha_j \cdot x_j + \sum_{i=1}^{m} \beta_i \cdot \left(\sum_{j=1}^{n} b_{ij} \cdot x_j + \sum_{k=1}^{m} s_{ik} \cdot z_k\right) = 0 \quad .$$

Collecting all the coefficients of x_j in the new coefficients $\tilde{\alpha}_j$ we can write

$$\sum_{j=1}^{n} \tilde{\alpha}_j \cdot x_j + \sum_{i=1}^{m} \beta_i \cdot \sum_{k=1}^{m} s_{ik} \cdot z_k = 0$$

or after interchanging the order of summation

$$\text{(A2)} \qquad \sum_{j=1}^{n} \tilde{\alpha}_j \cdot x_j + \sum_{k=1}^{m} \left(\sum_{i=1}^{m} \beta_i \cdot s_{ik}\right) \cdot z_k = 0 \quad .$$

Now, if S did not have maximal rank, the columns of S would be linearly dependent. This is equivalent to

$$\sum_{i=1}^{m} \beta_i \cdot s_{ik} = 0 , \quad k = 1,2,\ldots m$$

even if *not* all $ß_i = 0$. So (A2) and (A1) would be fullfilled with at least one $ß_i \neq 0$ which is in contradiction to the positive definiteness of fBm.

6 References

[FEL85] A. Fellous, J. Granara, J.C. Hourcade, Fractional Brownian Relief: An Exact Local Method, in: Proceedings of Eurographics 85 (North-Holland, 1985), pp. 353 - 363.

[FOU82] A. Fournier, D. Fussell, L. Carpenter, Computer Rendering of Stochastic Models, Comm. of the ACM 25 (1982), pp. 371 - 384.

[LAM77] J. Lamperti, Stochastic Processes (Springer, 1977)

[LEW87] J.P. Lewis, Generalized Stochastic Subdivision, ACM Transactions on Graphics 6 (1987), pp. 167 - 190.

[MAN82] B. Mandelbrot, The Fractal Geometry of Nature (Freeman, 1982)

[RÜM90] W. Rümelin, Simulation of Fractional Brownian Motion, in H.-O. Peitgen, J.M. Henriques, L.F. Penedo (eds), FRACTAL 90 - Proceedings of the 1st IFIP Conference on Fractals, Lisbon, June 6 - 8, 1990 (Elsevier, to appear)

[VOS88] R.F. Voss, Fractals in Nature: From Characterization to Simulation, in: H.-O. Peitgen, D. Saupe (eds.), The Science of Fractal Images (Springer, 1988), pp. 21 - 70.

Figure 3: Input data for the interpolation algorithm: 13 x 13 data points of the *Matterhorn* with meshsize 250 m

Figure 4: The *Matterhorn* generated by fractal interpolation of the data shown in figure 3 with fractal dimension 2.3: 385 x 385 points with meshsize 7.8 m

Figure 5: The *Matterhorn* generated by fractal interpolation of 54 irregularly spaced data points with fractal dimension 2.3: 401 x 401 points with meshsize 7.5 m

Plate 1 and 2
see page 194, figures 5 and 6

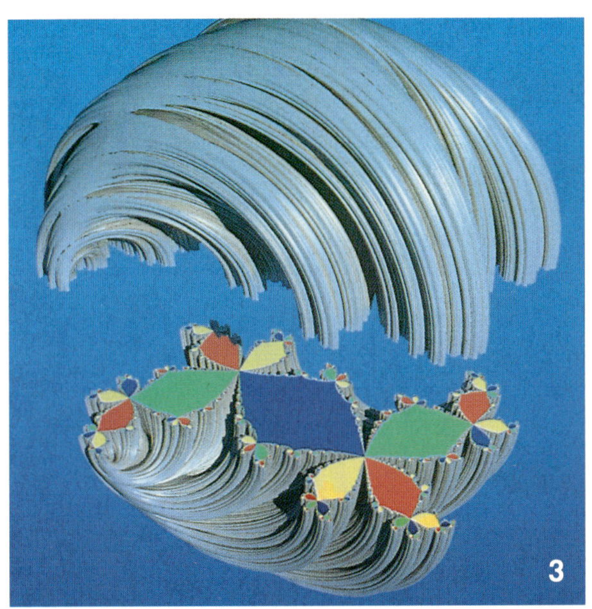

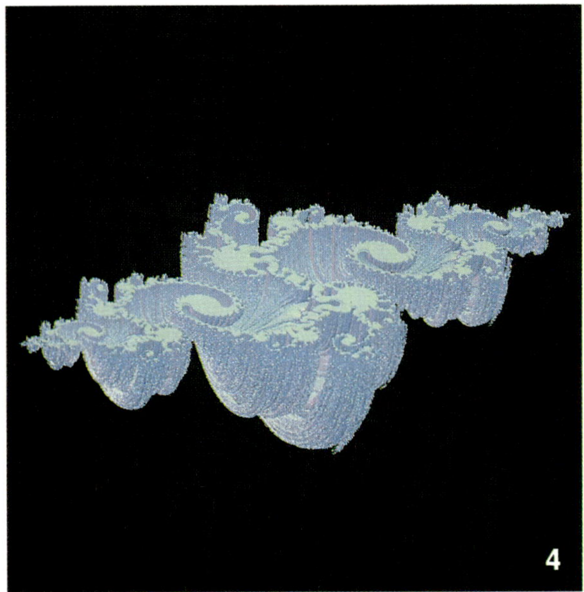

Plate 3
see page 109, figure 4a

Plate 4
see page 108, figure 3a

Plate 5
see page 115, figure 3

Plate 6
see page 119, figure 6

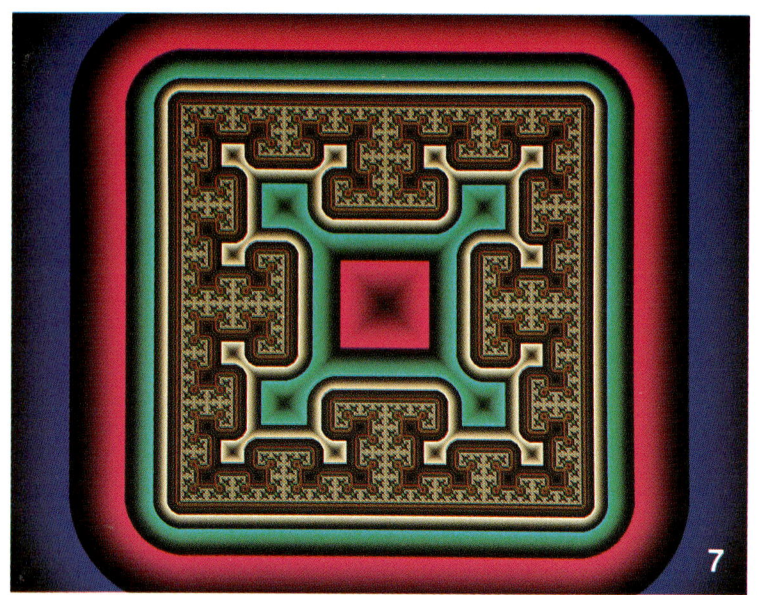

Plate 7
see page 43, Plate 1

Plate 8
see page 43, Plate 3

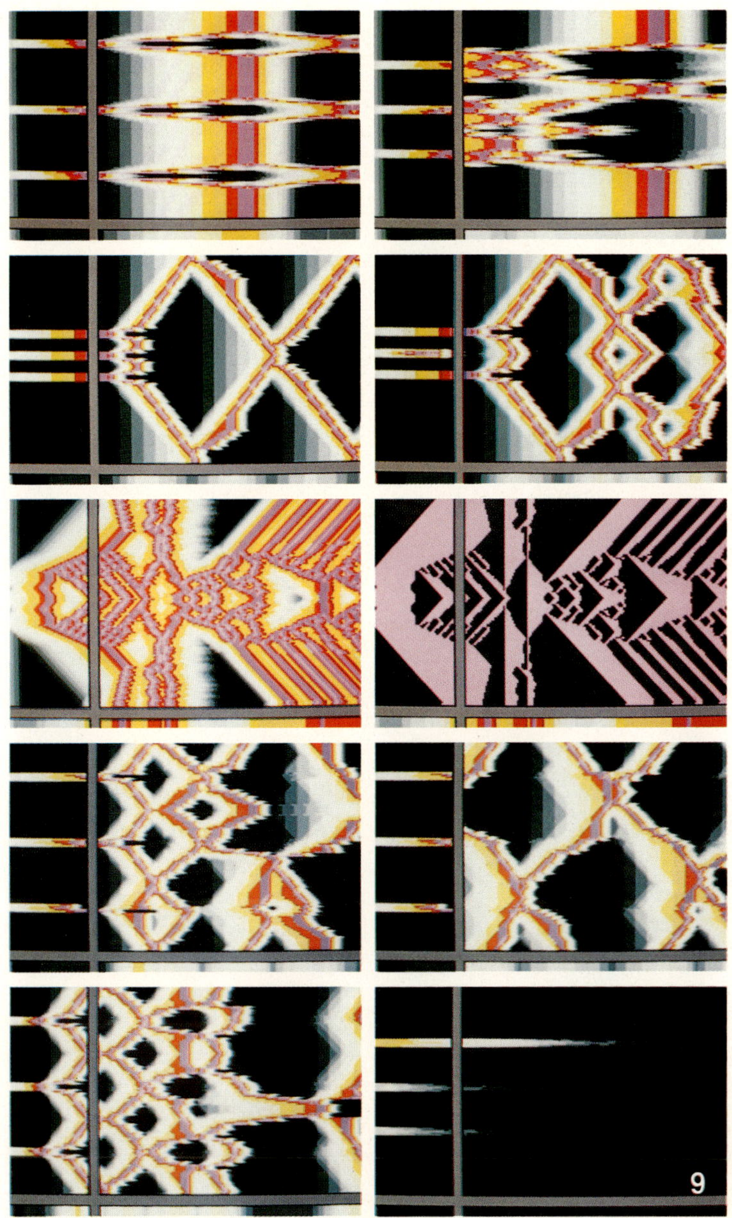

Plate 9
see pages 153–160, figures 2–11

Plate 10

Plate 11
see page 44, figure 14

Plate 12
see page 170, figure 9

III. Part: Simulation

Simulation of Malignant Cell Growth

W. Düchting

Department of Electrical Engineering, University of Siegen

Abstract: The aim of this paper is to show how systems analysis, control theory and computer science can stimulate new approaches to interpret cancer as a structural unstable closed-loop control circuit, to simulate temporal and spatial tumor growth, and to optimize cancer treatment by computer simulation.

1 Biological Observations

Cancer is a multistep process with the stages of initiation, promotion and progression. Characteristic features of malignant tumors are uncontrolled proliferation, invasion in adjacent normal tissue, metastases induced to other tissues via lymphatic channels, and the ability to evade immune surveillance. Recent research activities have focused on the field of molecular biology especially on oncogenes and suppressor-genes [WEI89].

In spite of this progress the main question how genes and the growth of normal and malignant cells are regulated still remains open.

Most of the normal tissues in the body contain some cells (liver cells, kidney cells) that can renew themselves if a tissue is injured. The division of a cell into two new ones involves four stages: G1 > S > G2 > M (G1: initial resting phase; S: the synthetic phase during which the doubling of DNA occurs; G2: a second resting phase or premitotic phase; M: the actual process of mitosis). When the replacement has been completed the repair process stops. Furthermore, at particular stages of the cell cycle the cells may move out of the cell cycle into a resting phase known as G0.

In contrast to the normal cell a tumor cell is theoretically able to divide indefinitely. In addition, a different morphology, larger nucleus, abnormal number of chromosomes and the formation of new capillaries (tumor angiogenesis) which is associated with a more rapidly growing tumor can be observed.

For studying the process of carcinogenesis tumors are induced to animals or to cell cultures (in vitro). Cell cultures are not only used to study the division of tumor cells, but also to determine the effect of chemotherapeutic drugs and of irradiation. During the past years a large progress has been made in cellkinetic experiments gaining hard data about normal and abnormal cell-growth control processes, for instance of cell-cycle phase durations.

Starting from basic biological test results a large body of mathematically oriented work applying mathematics to the field of biology and medicine has been published [WHE88]. Unfortunately, these models which consist of complicated formulae, are in most cases not completely understood by clinicians. In this dilemma the combined application of methods of systems analysis, control theory, automata theory, computer sciences and heuristics is a good link between the diverging areas of medicine and mathematics.

2 Design Strategy of Cancer Modeling

Modeling cancer may be performed at different levels (molecular level, cellular level, organ level). In our approach we focus on modeling at the cellular level because at present our experimentally gained knowledge about the control mechanisms at the gene level is still very poor and diffuse.

When constructing a computer model of a biological system we have to decompose a complex system into several sub-systems. For this top-down design the construction of a model describing cancer growth requires:

- cytokinetic models which describe the cell division of normal and tumor cells at a cellular level including experimentally gained data e.g. of cell-cycle phase durations.
- heuristic cell-production and interaction rules describing the cell-to-cell communication. For instance one rule of the catalogue may say: All cancer cells residing at a distance larger than 100 μm from the capillaries after the next division step will enter the resting phase G0.
- transport equations (diffusion-, Poisson-equation) describing cell movement, that means we have to introduce gradients of pressure and metabolic compounds into the model.
- computergraphics software packages for representing 2D and 3D simulation results.
- powerful and fast computers.

The large body of statements, rules and equations is transformed into algorithms. In addition, algorithms considering cancer treatment (surgery, chemotherapy and ratiation therapy) are developed in subprograms written in FORTRAN IV. To start the simulation program the following input data have to be fed into the computer: Notations about the character of a cell (normal, malignant), cell-cycle phase durations, cell-loss rates, initial configuration of normal tissue and of tumor cells and distinguished data about the kind of the planned cancer treatment.

3 Simulation of Malignant Cell Growth

Our approach developing closed-loop control circuits for cancer growth started in 1968. At that time the subject of consideration was focused on stability conditions and on the key idea of interpreting cancer as an unstable closed-loop control circuit [DUE68]. A very first application investigated the dynamic behavior of the formation of red blood cells. The outcome of this simulation experiment was the number of erythrocytes as a function of time. Thus, it was possible to interpret different blood diseases by one and the same multi-loop control model.

Then, oncologists advised us to study not only the dynamic behavior of cancer cells in the time domain, but also in the space. Starting with modeling the 2D multiplication of a normal cell inoculated into a nutrient medium (Petri dish) we tried to simulate cancer growth in the tissue of a tobacco leaf [DUE80], which is demonstrated in *Figure 1*. In this model we can only distinguish between normal and malignant cells, but we cannot say in which phase of the cell cycle each individual cell is residing.

In the next step the stream of ideas made us introduce distinguished cell-cycle phases (G1, S, G2, M, G0, N) for each cell. Thus, we were able to simulate 3D growth of a single dividing cancer cell [DUE81], inoculated into the center of nutrient medium at the beginning of the simulation run. *Figure 2* shows the steady state of a tumor spheroid in vitro. One can clearly recognize the balance between the proliferating tumor cells in the outer viable rim and the inner necrotic zone.

It is near at hand to extend the model developed so far from in-vitro to in-vivo cancer growth. The tremendous problem which had to be solved was the substitution of the nutrient medium by a capillary network. In [VOG86] Vogelsaenger made a simplified approach of modeling and simulating the formation of capillaries during the ontogenesis of a brain segment of a rat. The rationale behind this effort is the description of the formation of blood vessels as a

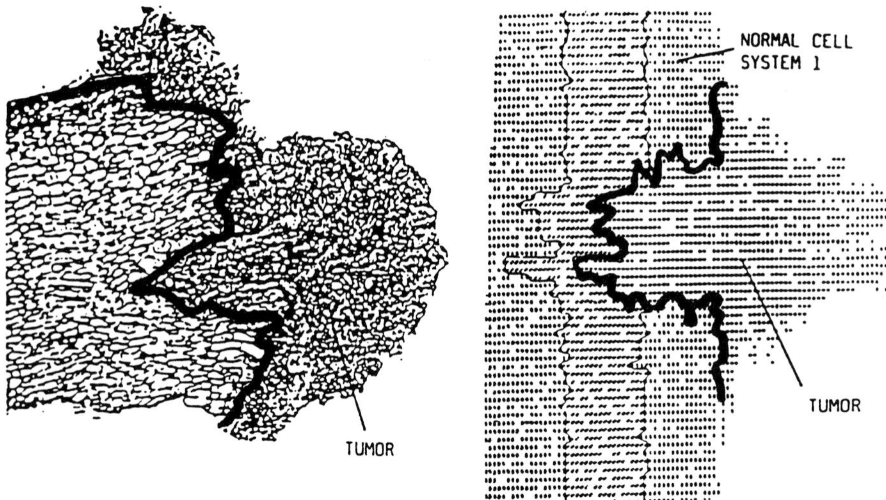

Fig. 1: Simulation of cancer growth in the tissue of a tobacco leaf

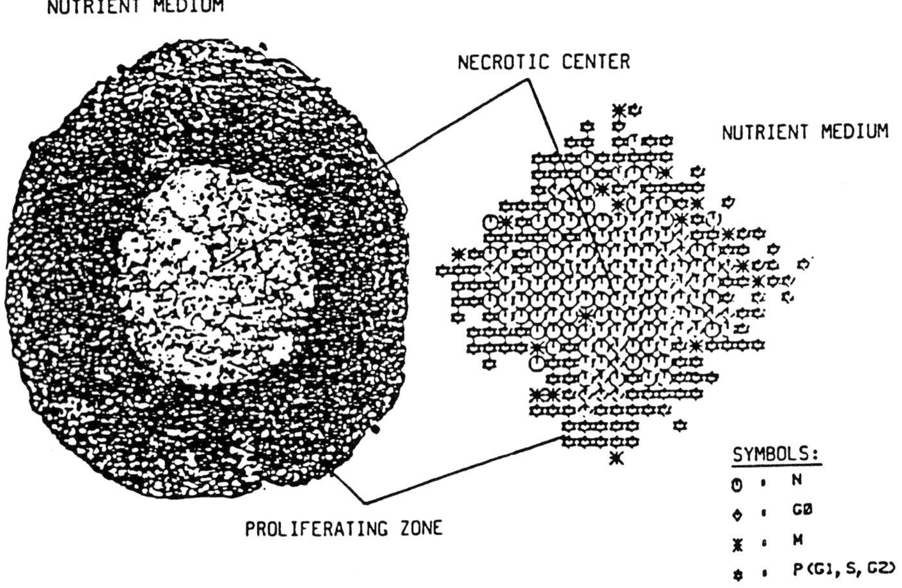

Fig. 2: Formation of a tumor spheroid (The initial configuration consisted of a single mitotic tumor cell placed in the center of the mutrient medium)

regulated process controlled by the demand for energy (0_2, Glucose) of each single cell. Then the assumption could be made that an individual cancer cell is arbitrarily placed in the normal tissue of the cortex of a rat at T=1 unit of time *(Figure 3)*. It turns our that the spread of cancer cells can be simulated representing the formation of a micrometastasis *(Figure 4)*.

4 Simulation of Cancer Treatment

If we want to model cancer treatment we have to extend the models of cancer growth by developing additional program packages describing the different treatment methods and schedules of surgery, chemotherapy and radiation therapy. A relatively simple simulation experiment is the surgical removal of a tumor in vitro and in vivo. In [DUE80] we have demonstrated the growth of a skin tumor *(Figure 5 (a)-(b))* which was only partially removed in *Figure 5 (c)*. Subsequently the remaining tumor cells continue to grow which is pointed out in *Figure 5 (d)*. The impact of simulating surgical treatment mainly lies in the field of education and training medical doctors.

The development of a radiation therapy model is much more complicated than that of a surgical model. In case of radiation therapy we started with the model of in-vitro tumor spheroids described in [DUE81]. To construct a model describing radiation treatment it is necessary to know the number of cancer cells hit by radiation. In our model [DUE89] we have made use of the survival function S(D) via the "Linear Quadratic Model" (LQM) which allows to compute the number of hit specific tumor cells as a function of the dose. According to the calculated number of the cells to be killed pseudo-random number generators perform this task in our model. In this way, it is possible to test different clinical irradiation schemes on in-vitro tumors. The question "What is more favourable, a multifractionated irradiation or an irradiation with a high single dose per week?" is in permanent controverse discussion. The unexpected simulation results in *Figure 6* and *Figure 7* demonstrate that in both cases, with nearly the same overall dose after five weeks, the number of tumor cells has decreased to about the same level. Therefore, one may speculate that the optimal irradiation scheme depends on the radiation response of normal cells including side-effects. Much work remains to be done in the future to model the complete scenario of irradiation of heterogeneous tissue including normal and cancer cells.

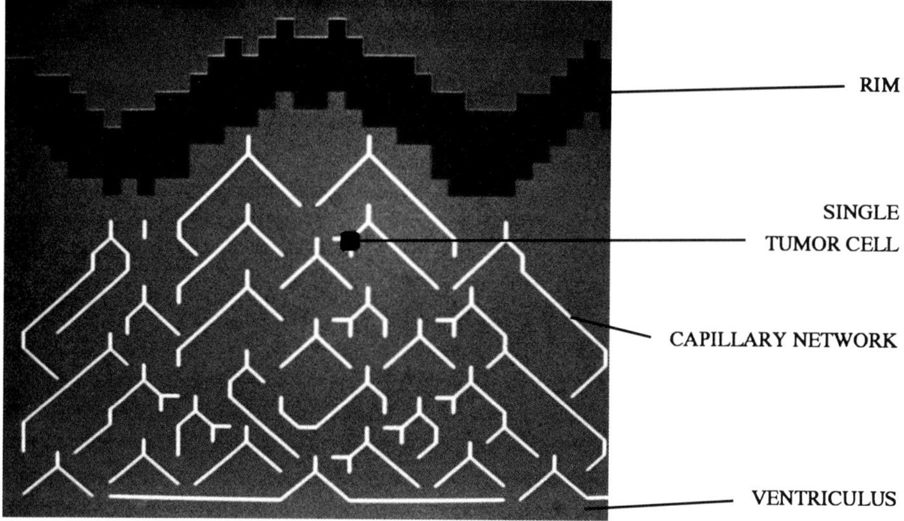

Fig. 3: A single tumor cell is introduced in the tissue of the cortex of a rat at T = 1 unit of time

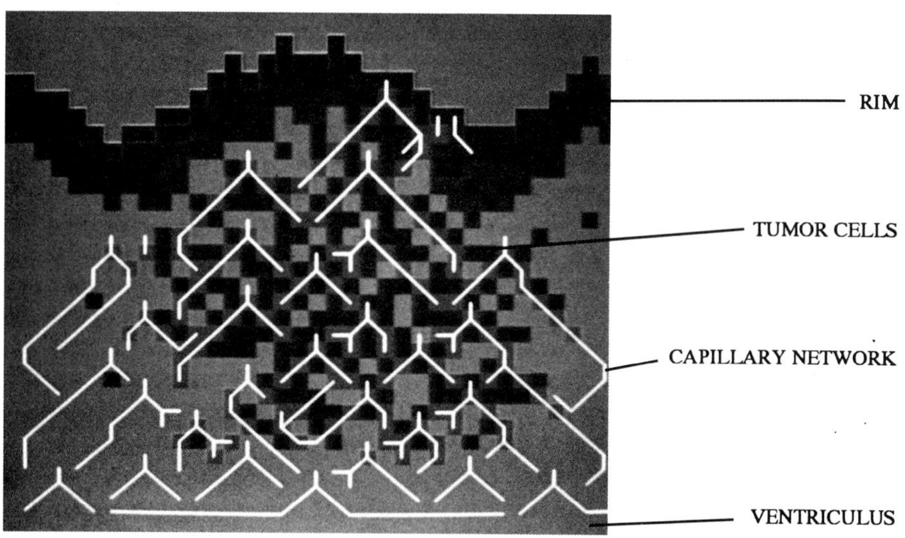

Fig. 4: Spread of tumor cells in the cortex of a rat at T = 100 units of time

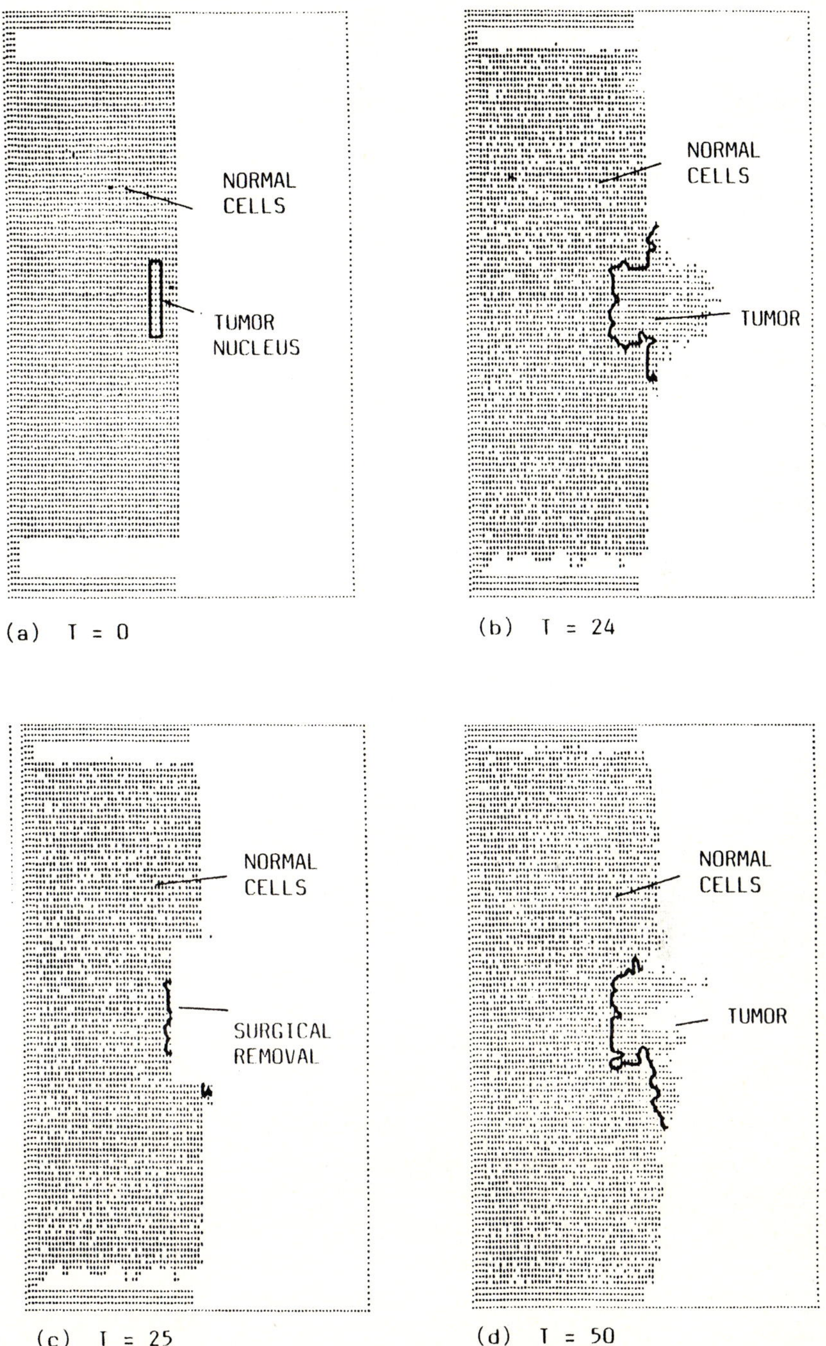

Fig. 5: Tumor growth (a), (b); surgical removal of tumor cells (c), and recidiv (d)

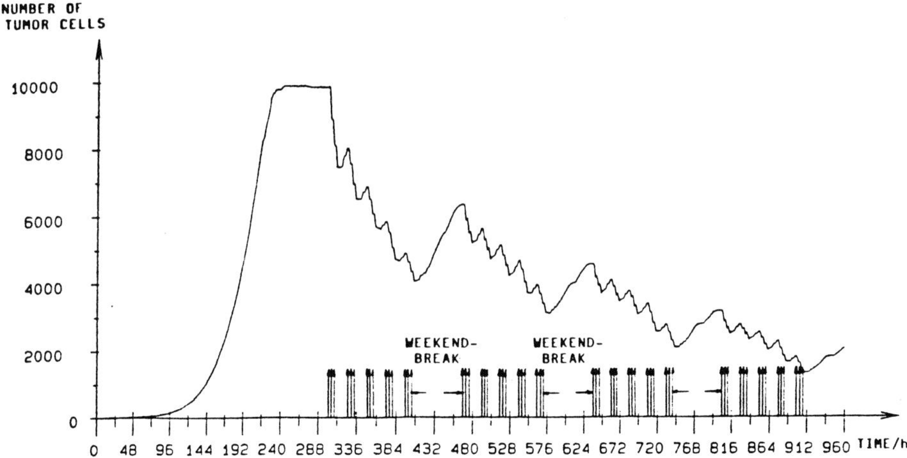

Fig. 6: Simulation of a multifractionated irradiation of a tumor spheroid (adenocarcinoma of the mouse): 5 x 3 x 0.7 Gy per week ; overall dose: 63 Gy; 30% of the hit cells will be repaired after 15 h; lysis duration of the lethally hit cells: 5 days

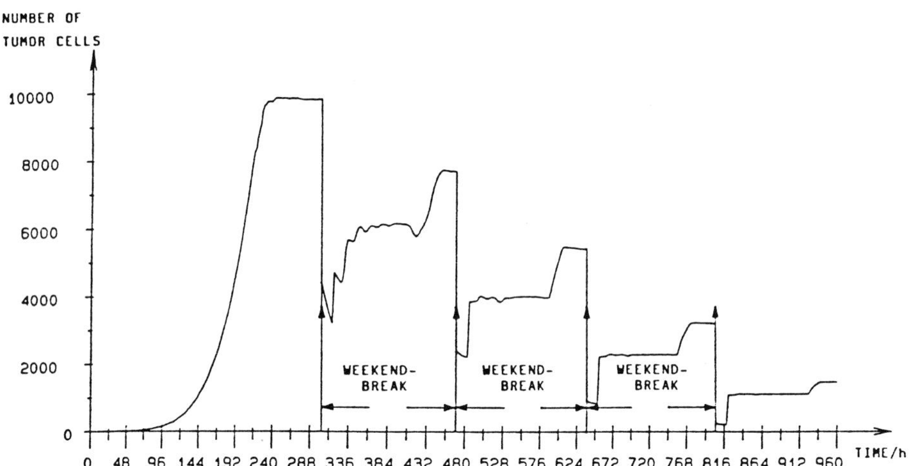

Fig. 7: Simulation of an irradiation of a tumor spheroid with a high single dose (adenocarcinoma of the mouse): 1 x 6 Gy per week; overall dose: 60 Gy; 30% of the hit cells will be repaired after 15 h; lysis duration of the lethally hit cells: 5 days

5 Open Questions

Factors not included in our models so far are: Heterogenity, immunologic reactions, drug resistance, formation of metastases and side effects. These are promising avenues of future research work. Furthermore, we are working in the field of combining models of tumor growth with image processing techniques (CT, NMR, PET) to provide a more realistic initial configuration for our treatment models.

6 Literature

[DUE68] W. Düchting: Krebs, ein instabiler Regelkreis, Versuch einer Systemanalyse, Kybernetik, 5. Band, 2. Heft (1968): 70-77

[DUE80] W. Düchting, G. Dehl: Spatial Structure of Tumor Growth: A Simulation Study, IEEE Transactions on Systems, Man and Cybernetics SMC-10, No. 6 (1980): 292-296

[DUE81] W. Düchting, T. Vogelsaenger: Three-Dimensional Pattern Generation applied to Spheroidal Tumor Growth in a Nutrient Medium, Int. J. Bio-Medical Computing 12 (1981): 377-392

[DUE89] W. Düchting, R. Lehrig, G. Rademacher, W. Ulmer: Computer Simulation of Clinical Irradiation Schemes Applied to In Vitro Tumor Spheroids, Strahlenther. Onkol. 165, Nr. 12 (1989): 873-878

[VOG86] T. Vogelsaenger: Modellbildung und Simulation von Regelungsmechanismen wachsender Blutgefäßstrukturen in normalen Geweben und malignen Tumoren, Dissertation Siegen, Siegen, 1986

[WEI89] R. A. Weinberg: Oncogenes and the Molecular Origins of Cancer, Cold Spring Harbor Laboratory Press, Cold Spring Harbor, 1989

[WHE88] T. E. Wheldon: Mathematical Models in Cancer Research, Adam Hilger, Bristol, 1988

Simulation of Individual Behaviour

E.J.Swart and P.J.Plath
Universität Bremen, Bremen

1 Introduction

It is becoming more and more apparent that social systems - especially human social systems - can be described by deterministic models [BAC90, BAK91, CAL89, WEH90]. There are examples of quite different types: technological and economic systems [SIL88, BAU89, EBE91, JIM80], behaviour in road traffic [KUE84] and also the global behaviour of people in states [LUH87, WEH90]. It is crucial that these deterministic models are based on average or expectation values of a large number of elementary entities. Within the scope of this view deviations from these values should be understood as fluctuations. In critical situations such fluctuations are, of course, able to affect the whole system and possibly to provide it with a new structure. Such fluctuations, however, are incompatible with a purely deterministic view. Fluctuations are random events and can therefore only be described probabilistically.

In the context of the theory of social systems, however, the idea of individual decisions able to influence the whole system plays an important rôle. The idea of freedom is essentially a theory based on individuals. The concept of democracy according to which decisions are made by means of elections, for instance, is based on individual freedom to take decisions, elections only being a certain form of standardisation of this behaviour. On the other hand, historical events are not represented by an enumeration of election results. They are, on the contrary, bounded to special persons, that is to say to selected individuals. So the question arises as to how to combine both views, the deterministic description of global behaviour and the historical description of individual actions.

In the following, the attempt is made to simulate - by means of cellular automata - the rôle of elementary entities able to take individual decisions within a system which otherwise is fully deterministic.

2 Basic Ideas behind the Model

If a system is essentially described deterministically, individual random decisions should be important with respect to the structure of the system if the units taking these decisions are rare enough. This concept differs fundamentally from the thermodynamic concept of the general presence of fluctuations, where all parts of the system are always able to take individual decisions. So we assume that our system can be described by an irreversible thermodynamics far from equilibrium, and that the behaviour of most of the parts of the system can be described fully deterministically. We are thus dealing with a system that is able to generate a structure, and which is far from equilibrium in a thermodynamic sense. Such a system can be subdivided into a large number of units which behave irreversibly but deterministically in time.

Into such a system we introduce only a few individualists. These units are characterised by the fact that at any time they can - with a given probability - randomly derive a little from the deterministically determinable behaviour, which also applies to them. We are interested in the effect of individual behaviour on the development of the whole system. The individualists may be considered as a source of steadily random disturbances with respect to the development of the system. If such a disturbance has no effect, or merely a temporary one on the total behaviour we would call such a system stable. If, however, the disturbance has a long-term structuring effect, the underlying deterministic system is unstable and a new structure is formed. Then, the new structure is essentially based on the effect produced by the individualists. It will be of great interest, whether in such a case the whole system is caught by the new structure, or whether the new structure is only of local importance covering just a relatively small part of the system.

We emphasise once more that we do not consider the deterministic system to be a system in equilibrium. We are therefore interested in the local or global stability or unstability of a structure which is not in equilibrium.

In our introduction we referred to human systems in particular. Nevertheless, we are convinced that our concept also applies to biological or chemical systems, if they are complex enough. This is always the case if we are regarding a system where most of the parts can be described solely in terms of their average values in a deterministic sense. Some systems, however, cannot be described by their average values, since these are in a very critical and unstable state. Such systems are to be described, therefore, in terms of fluctuations around the development of their average values. It would appear that these exceptional, unstable and strongly varying subsystems can structurally influence the whole system or parts of it under certain circumstances.

As Luhmann [LUH87] explains, a social system is a system which essentially only concerns itself and has barely any contact to its outer world. So the system creates for its own parts an internal environment reflecting the behaviour of the whole system with respect to its outer environment. This internal world can be decomposed into two components: the local environment of each unit - usually described by means of a cellular automaton - and a global internal environment representing the whole system, which can be described by the average value of any component of the system. In the following, we provide two examples for the effect of both internal representations of the outer world:

- systems which destroy or strongly affect the conditions of their existence due to the way in which they produce or
- systems creating conditions which favour their development and existence.

In our model we shall represent this global influence of the system on the individual processes of the single units by the average value of the productivity of all cells.

The rare antennae of perception of the outer world of the system that Luhmann speaks about, are translated here into the internal representation of the system in terms of the system-immanent representation of the outer world. In this sense our individualists and their random decisions operate as internal representatives of the outer world.

In our model, the random decisions of the individualists only concern their productivity. Each cell is, however, as already mentioned, also characterized by a further component: a phase. The temporal transformation of the phase of each cell always takes place deterministically under consideration of the local and global properties of the system. While the productivity describes the quantitative behaviour, the phase describes a qualitative property of the cell. In a certain way the phase represents a short-lived structural memory of the single cells.

The interpretation of the terms productivity and phase depends on the actual system to be modelled. With regard to the idea of productivity we call the states of the phase the active or inactive behaviour of the cell. From a thermodynamic point of view the idea of the phase is a structural property of the cell, which is a subsystem, so the term phase is used in a slightly different way here. We assume that a transition can take place between both discrete values of the phase. In a way similar to the productivity, the temporal transformation of the phase depends on the productivity and phase of the neighbouring cells and on the global system.

While dynamic systems are usually characterized by the stability or instability of their state, for social organisms another aspect appears, which cannot be described by the Lyapunow stability criterion. It is the ability of the system to protect itself against overexcitement as well as against dying off because of a lack of stimuli. Thresholds slow down increase and decrease of the productivity and the exceeding of upper and lower limits is avoided, where this could end the existence of the system.

Moreover, we presume that social organisms are able to stimulate themselves when their productivity has decreased long enough or deep enough. This fact is modelled by a phase transition into the active state, although - based on the local situation - the cell had to stay in an inactive state. This rule is the result of the global self-control of the system. It may be understood as the self-stimulus of the system.

In the following paragraphs we shall present the formal mathematical description of the model in detail. We shall then describe some special situations by way of example.

3 Cellular Automata

In the following we briefly explain our concept of cellular automata [WOL84, TOF87].

- A cellular automaton is a set of cells with a structure. The structure on the set is a simplicial complex of dimension one, where the cell is a vertex.
- Each cell is labelled by a state, which may be a scalar, a vector, etc.
- A local neighbourhood between the cells is defined.
- A transformation rule transforms the state of a cell at the time t into the state of the cell at the time $t+1$ dependent on the state of the cell(s) in the defined neighbourhood.

A simple example may illustrate our concept:

- A set of cells is ordered like a chain.
- The state of the cells is a binary value; "true" or "false".
- Every cell - except the first one - has its predecessor as a neighbour. The first cell has no neighbour and its state is "false" at the time t_0.
- The state of the other cells is transformed by the exor-rule: $c_i(t+1) = c_i(t)$ exor $c_{i+1}(t)$.

This rule generates a Sierpinsky pattern.

The cellular automaton to be presented here follows the concept described above.

- The automaton consists of a set of 128 cells ordered as a chain, which may be closed to form a circle.
- The state of a cell is described by a vector with three components.
- Every cell has its predecessor and its successor as its neighbours. If the automaton is not closed to a circle, there are two possibilities for treating the missing neighbours of the first and the last cell.
- There exists a set of transformation rules.

Figures 2 - 11 show examples of patterns generated by the rules of our system.

3.1 A Cell's State

As already mentioned, the state of a cell is described by a vector with three components:

phase
productivity
individuality

- The phase of a cell represents the qualitative behaviour of an element of the system to be modelled. It is binary: At a time t a cell c_i may be activated (value 1) or not (value 0):

$$\text{phase}(c_i(t)) \in \{0, 1\}$$

- The productivity represents the quantitative behaviour of an element of the system to be modelled. It is expressed in numbers from zero to 111:

$$\text{prod}(c_i(t)) \in \{0, .., 111\}$$

- A cell is either an individualist (value 1) or not (value 0). Whether a cell has this property or not is fixed before starting and not changed during the development of the automaton.

$$\text{indiv}(c_i(t)) \in \{0, 1\}$$

Figure 7 shows an example of the development of the phases, while examples of the development of the productivity of the cells are shown by Figures 2-6 and 8-11.

3.2 The Cell's Neighbourhood

In principle, two neighbours are defined for each cell: its predecessor and its successor. If the chain of cells is closed to a circle, there is no exception to this rule. Otherwise, the first and the last cell only have one neighbour: a successor and a predecessor respectively. The missing neighbours can be replaced in two different ways:

- They are considered to be inactivated and their productivity is equal to zero.

- Their phase and their productivity are considered to be equal to the phase and the productivity of the first and the last cell respectively.

3.3 The Set of Transformation Rules

At time t the state of a cell is described by a vector with three components:

$$state(c_i(t)) = \begin{pmatrix} phase(c_i(t)) \\ prod(c_i(t)) \\ indiv(c_i(t)) \end{pmatrix}$$

With each time step the phase and the productivity of a cell c_i are calculated anew.

3.3.1 Transformation of the Phase

The phase of the cell c_i at time $t+1$ is calculated by means of the following formula:

$$c_i(t+1) = 1, case((a \geq \acute{a}) \wedge (b < \acute{b})) \vee (s < \acute{s})$$
$$c_i(t+1) = 0, otherwise$$

where

- $a = (phase(c_{i-1}(t)) + phase(c_i(t)) + phase(c_{i+1}(t)))$
  $threshold\ \acute{a} \in \{0, .., 2\}$
- $b = (prod(c_{i-1}(t)) + wgt \times prod(c_i(t)) + prod(c_{i+1}(t))) \div (wgt + 2)$
  $wgt \in \{1, .., 9\}$
  $threshold\ \acute{b} \in \{1, .., 111\}$
- $s = \sum_j phase(c_j(t)), (j = 1, .., 128)$
  $threshold\ \acute{s} \in \{1, .., 128\}$

So the phase of a cell at the next time step is a functional F described below:

$$\begin{aligned} phase(c_i(t+1)) = F(\quad & f_1(\ phase(c_i(t)), \\ & \quad phase(c_{i-1}(t)), \\ & \quad phase(c_{i+1}(t)), \\ & \quad threshold\ \ \acute{a}), \\ & f_2(\ prod(c_i(t)), \\ & \quad prod(c_{i-1}(t)), \\ & \quad prod(c_{i+1}(t)), \\ & \quad weight\ \ wgt, \\ & \quad threshold\ \ \acute{b}), \\ & f_3(\ \sum_j phase(c_j(t)), (j = 1, ..128) \\ & \quad threshold\ \ \acute{s})) \end{aligned}$$

This means that calculation of the cell's phase at time $t+1$ depends on

- the phase of the cell and the phases of its neighbours,
- the weighted productivity of the cell in question and the productivity of its neighbours,
- the number of activated cells in the automaton,
- the thresholds $\acute{a}, \acute{b}$ and $\acute{s}$.

Thus there are several possible ways to influence the temporal development of the automaton with respect to the phase of the cells, namely by varying

- the threshold for the sum of the phases of the actual cell and its neighbours,
- the threshold for the average productivity of the actual cell and its neighbours,
- the weight of the productivity of the actual cell,
- the threshold for the number of activated cells.

3.3.2 Transformation of the Productivity

The increase or decrease of the productivity of a cell c_i during the transformation from t to $t+1$ is calculated by means of the following formula:

$$\begin{array}{lcll}
prod(c_i(t+1)) & = & prod(c_i(t)) - pd1, & phase(c_i(t)) = 0 \\
prod(c_i(t+1)) & = & prod(c_i(t)) + pd2, & phase(c_i(t)) = 1 \\
 & & & \wedge d < \acute{d} \\
 & & & \wedge prod(c_i(t)) < b \\
prod(c_i(t+1)) & = & prod(c_i(t)) + pd3, & phase(c_i(t)) = 1 \\
 & & & \wedge d < \acute{d} \\
 & & & \wedge prod(c_i(t)) \geq b \\
prod(c_i(t+1)) & = & prod(c_i(t)) - pd4, & phase(c_i(t)) = 1 \\
 & & & \wedge d \geq \acute{d}
\end{array}$$

where

- $d = \sum_j prod(c_j(t)) \div 128, j = 1, .., 128$
  $threshold\ \acute{d} \in \{1, .., 111\}$
- $b = (prod(c_{i-1}(t)) + wgt \times prod(c_i(t)) + prod(c_{i+1}(t))) \div (wgt + 2)$
  $wgt \in \{0, .., 9\}$
- $pd1 \in \{0, .., 9\}$ $pd2 \in \{0, ..9\}$ $pd3 \in \{0, ..9\}$ $pd4 \in \{0, .., 9\}$
- if the result of the decrease of productivity is < 0 $prod(c_i(t+1)) = 0$
- if the result of the increase of productivity is > 111 $prod(c_i(t+1)) = 111$

The productivity of a cell c_i at time $t+1$ is a functional G described below:

$$\begin{array}{lll}
prod(c_i(t+1)) = G(& g_1(& phase(c_i(t))), \\
 & g_2(& phase(c_i(t)), \\
 & & \sum_j prod(c_j(t)) \div 128, \\
 & & j = 1, .., 128 \\
 & & prod(c_{i-1}(t)), \\
 & & prod(c_i(t)), \\
 & & prod(c_{i+1}(t)),
\end{array}$$

$$
\begin{aligned}
& \qquad \textit{weight} \quad wgt, \\
& \qquad \textit{threshold} \quad \acute{d}), \\
g_3(& \ \textit{phase}(c_i(t)), \\
& \ \sum_j \textit{prod}(c_j(t)) \div 128, \\
& \ j = 1, .., 128 \\
& \ \textit{threshold} \quad \acute{d}))
\end{aligned}
$$

The calculation of the increase or decrease of productivity of $c_i(t)$ to $c_i(t+1)$, therefore, depends on:

- the phase of the actual cell,
- the (weighted) productivity of the actual cell and the productivity of its neighbours,
- the average productivity of the automaton at time t,
- the threshold $\acute{d}$.

Thus the development of the productivity of the single cells and of the automaton can be influenced by varying

- the threshold for the average productivity of the automaton,
- the weight of the productivity of the actual cell,
- the amounts to be added to or subtracted from the productivity of the actual cell at time t in accordance with the rule described above.

3.3.3 Individual Decisions

So far, the development of the state of the cells is fully deterministic. If a cell is an individualist, however, it is able to change the amount of the increase or decrease of productivity by means of an individual decision. The individualist can decide to *diminish* this deterministic amount by one or two. It can also decide to *increase* the deterministic amount by one or two. It can even decide to *accept* the amount dictated by the transformation rule. The individual decisions are taken randomly, based on a probability distribution for the five different decisions, which can be justified in ten equal steps for each decision.

Fig. 1 shows the probability table for the individual decisions.

The symbols can be understood as follows:

$$
\begin{aligned}
<< \; &= \textit{the difference of productivity is decreased by } 2 \\
< \; &= \textit{the difference of productivity is decreased by } 1 \\
= \; &= \textit{the difference of productivity is not changed} \\
> \; &= \textit{the difference of productivity is increased by } 1 \\
>> \; &= \textit{the difference of productivity is increased by } 2
\end{aligned}
$$

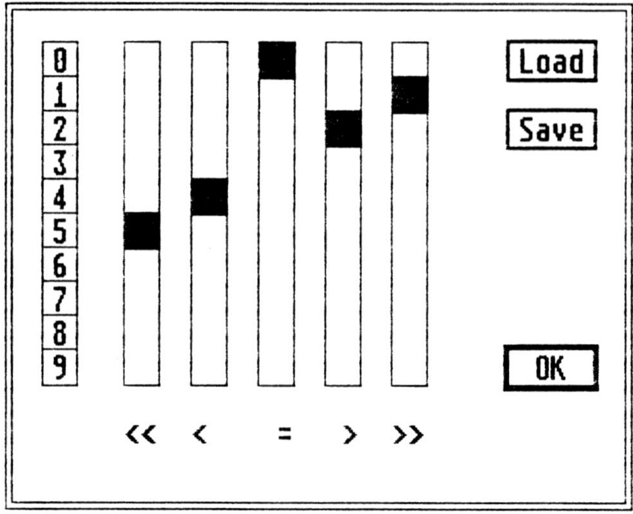

Figure 1: Probability table for individual decisions

4 Description of the Pictures

It is obvious that the rules of our system are of high complexity. We shall, therefore, not be able to describe each possible behaviour of the model in detail. We shall describe the characteristic behaviour of some particular automata instead. We hope that these examples can illustrate our initial general remarks, although in this context we can discuss them only roughly.

Rules for the automaton of Fig. 2:

1. probability distribution for individual decisions: 1/2/4/2/1 (normal distribution)
2. thresholds: $\acute{a} = 2$ $\acute{b} = 100$ $\acute{d} = 100$ $\acute{s} = 10$
3. weight $= 4$
4. difference of productivity: $pd1 = 3$ $pd2 = 3$ $pd3 = 1$ $pd4 = 1$
5. initialisation:
   (a) 3×5 active cells
   (b) 3×5 individualists
   (c) productivity of all cells $= 40$

The automaton has been initialised with three clusters each comprising five activated cells, all of them with the property to take individual decisions. Without them the automaton would grow into full productivity after a few time steps

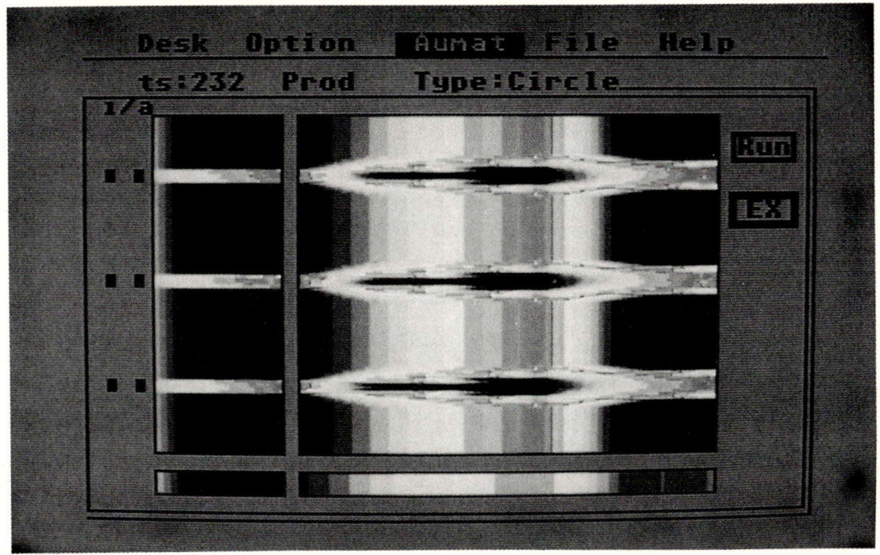

Figure 2: Isolated structuring by clusters of (initially) activated individualists

(violet colouring). The presence at the start of a cluster of active cells which are not individualists would not change the development essentially. The presence of one single individualist could not really change the state either. But providing two or more active cells on the periphery of the cluster with the property to be an individualist makes it possible to generate a structure. So we can conclude that the pattern we see here was influenced by these individualists, both in the horizontal structures, which are caused by local couplings, and in the vertical structures, which are caused by the global coupling of the system.

The state of maximal productivity would be the attractive state in the absence of individualists in an active cluster. The state the picture shows, however, is the attractive state of the system caused by the influence of the individualists.

Rules for the automaton of Fig. 3:

1. probability distribution for individual decisions: as Fig.2
2. thresholds: as Fig.2
3. weight: as Fig.2
4. difference of productivity: as Fig. 2
5. initialisation: as Fig.2, but with a smaller distance between the clusters

If the distance between the clusters of individual cells is small enough, the local patterns can fuse after more or less time steps. In this case (Fig.3) we can observe that structures induced by isolated clusters later fuse to form a common structure after approx. 2000 timesteps. Depending on the rules, we observe a

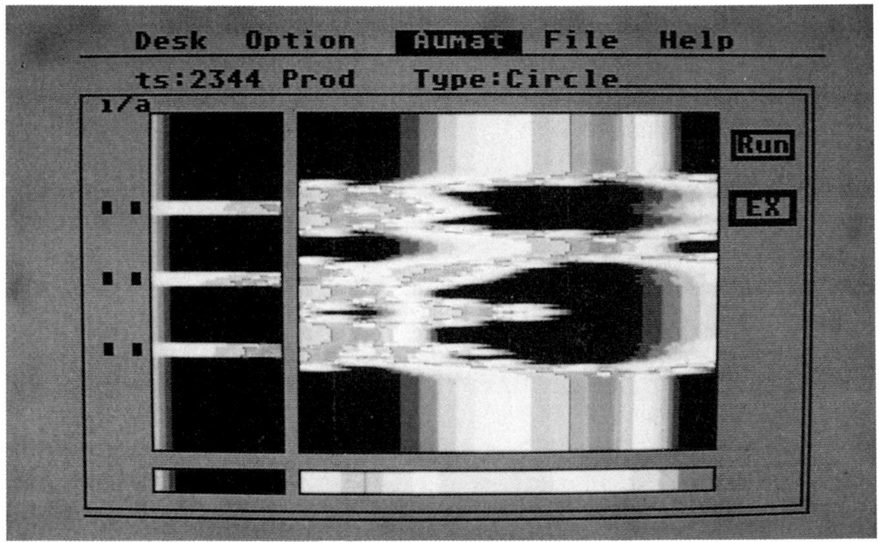

Figure 3: Fusion of structures caused by clusters of individualists with a critical distance

critical distance between the clusters of individualists for such a development.

Rules for the automaton of Fig. 4:

1. thresholds: as Fig.2
2. weight $= 2$
3. difference of productivity: $pd1 = 9 \quad pd2 = 9 \quad pd3 = 1 \quad pd4 = 1$
4. initialisation:
   (a) 3×5 active cells
   (b) no individualists
   (c) productivity of all cells $= 40$

Here we see the purely deterministic development of the automaton to an attractor. There are no individualists, so there is no random influence in this development. The structure however, is very unstable, as Fig.5 shows us.

Rules for the automaton of Fig. 5:

1. probability distribution for individual decisions: 1/2/3/4/5 lightly progressive
2. thresholds: as Fig.4
3. weight: as Fig.4

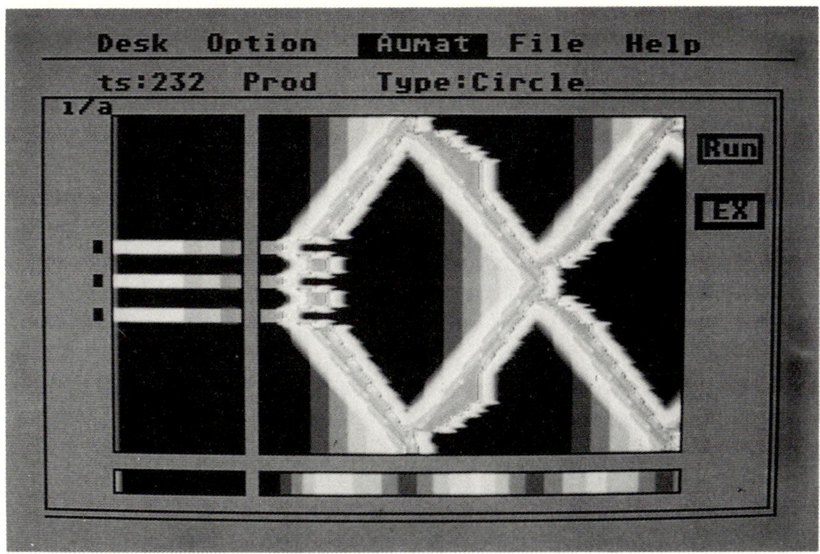

Figure 4: Deterministic Development

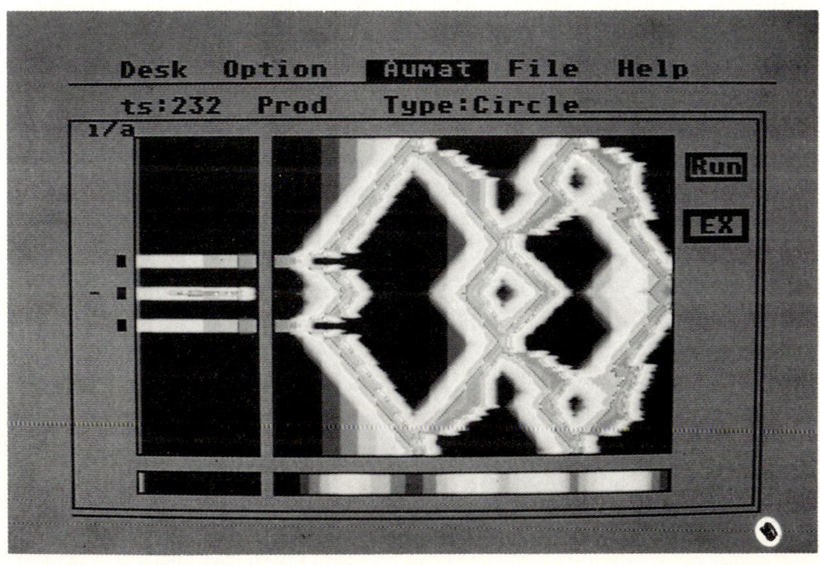

Figure 5: Disturbance of the structure in Fig.4 by one (!) individualist

4. difference of productivity: as Fig. 4
5. initialisation:
   (a) active cells as in Fig.4
   (b) one individualist
   (c) productivity of all cells = 40

The same automaton as in Fig. 4, but one of the active cells is now an individualist. As we can see, the original structure can already be transformed to a totally different structure by this one individualist alone, so, as we said already, the system without individualists (Fig.4) was very unstable.

Rules for the automaton of Fig. 6:

1. probability distribution for individual decisions: as Fig.5
2. thresholds: $\acute{a} = 1$, for the rest as Fig.2
3. weight: as Fig.5
4. difference of productivity: $pd1 = 3$ $pd2 = 9$ $pd3 = 2$ $pd4 = 2$
5. initialisation: as Fig.5

The fractal structure within limiting wave fronts has mainly been caused by changing the local neighbourhood relation with respect to the previous example. We changed the threshold $\acute{a}$ from two to one. This means that the chances for a cell to be in an active state at the next time step have increased, because only one of the three neighbouring cells under consideration has to be active for this purpose. Of course, it is still also necessary that the average productivity of the three neighbours does not reach or exceed the threshold $\acute{b}$.

Figure 7 shows the development of the phases of the same automaton.

To illustrate the differences between progressive, conservative (normal distribution) and regressive behaviour of the individualists we refer to the following three pictures:

Rules for the automaton of Fig. 8/9:

1. probability distribution for individual decisions: 1/2/4/2/1 (conservative behaviour)
2. thresholds: as Fig.2
3. weight: as Fig.2
4. difference of productivity: as Fig. 4
5. initialisation: as Fig.2

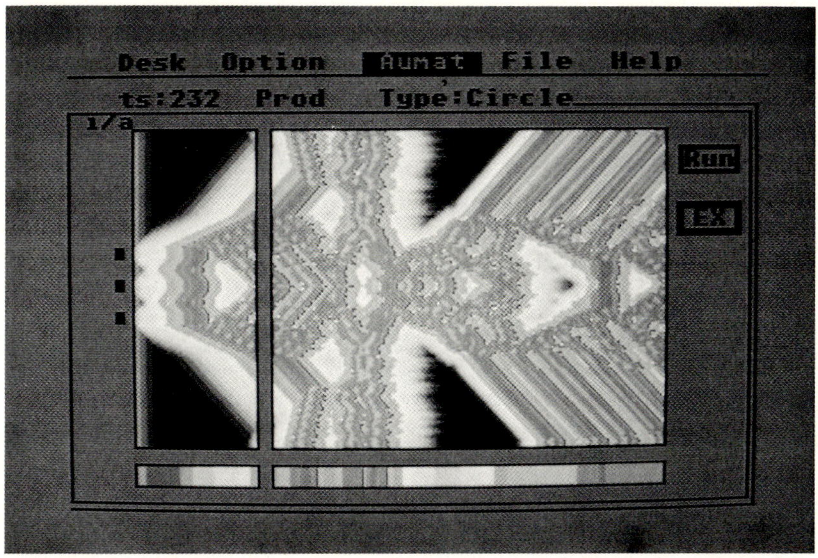

Figure 6: Interesting fractal structure of a purely deterministic automaton

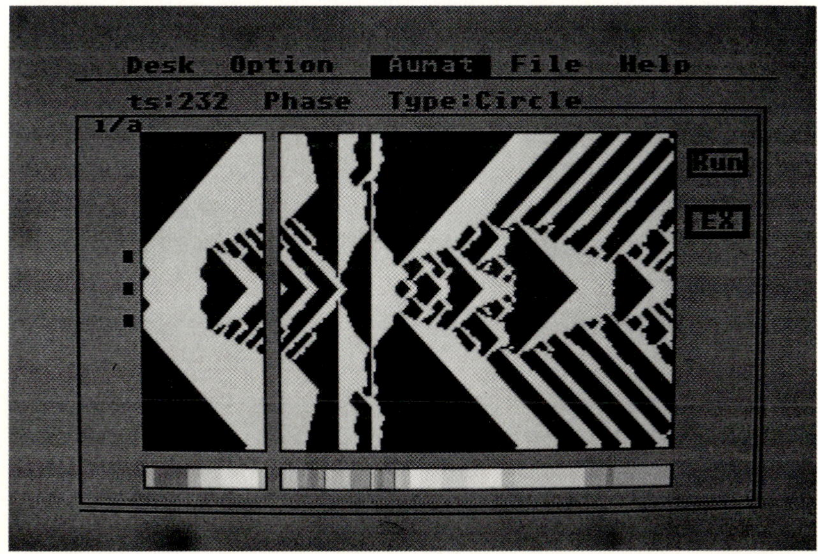

Figure 7: Development of the phases of the automaton of Fig.6

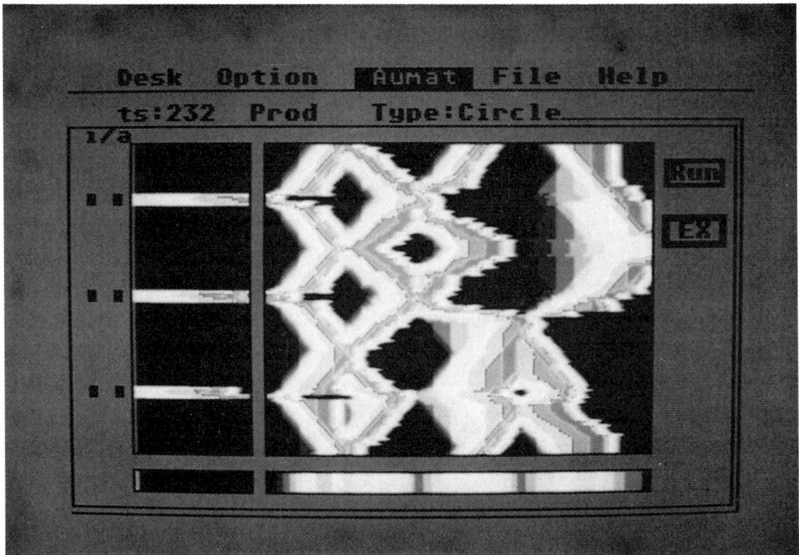

Figure 8: Start of the automaton with a conservative behaviour of the individualists

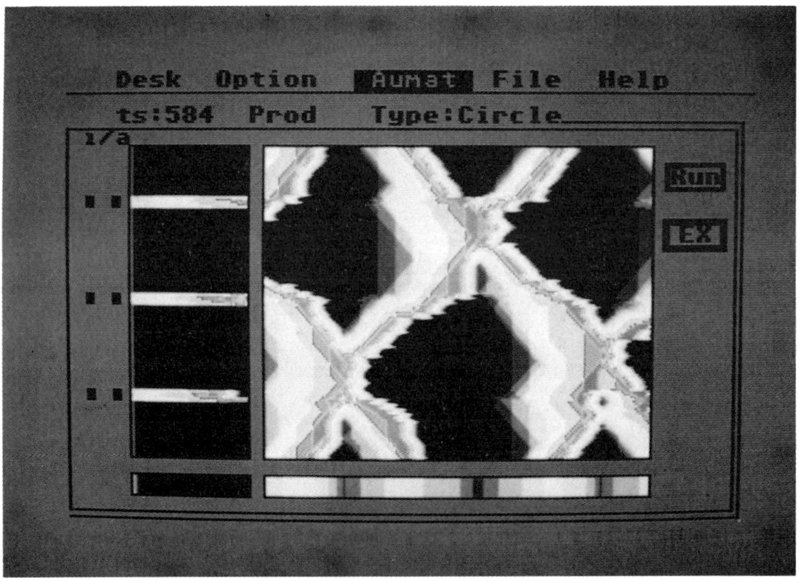

Figure 9: The conservative behaviour of the individualists of Fig.8 leads to a periodic development

The automaton develops a periodical behaviour.

Rules for the automaton of Fig. 10:

1. probability distribution for individual decisions: 0/0/0/7/9 (progressive behaviour)
2. thresholds: as Fig.8
3. weight: as Fig.8
4. difference of productivity: as Fig. 8
5. initialisation: as Fig.8

With exclusively progressive decisions of the individualists the periodicity which has been generated by the conservative automaton is strongly disturbed. The behaviour of the system becomes very chaotic.

Rules for the automaton of Fig. 11:

1. probability distribution for individual decisions: 9/7/0/0/0 (regressive behaviour)
2. thresholds: as Fig.8
3. weight: as Fig.8
4. difference of productivity: as Fig. 8
5. initialisation: as Fig.8, but the cells of the upper cluster start with a productivity of 96 (red colouring), while the other cells start with a productivity of 40 (as in all other examples).

With exclusively regressive decisions of the individualists, the productivity of the cells rapidly decreases (although the upper individualists started with a productivity of 96) so that the automaton goes down to the lowest level of productivity.

5 Conclusive Remarks

Of course, we are aware that social systems, especially such systems in which individual decisions are relevant, cannot be described totally by linear cellular automata. But some essential elements in the behaviour of a social system can surely be made visible by such a simple automaton. So, for instance, the question arises as to whether there are situations where such individualists are able to change the structure of a social system even by random behaviour. We believe we have given a positive answer to this question.

We have not examined exactly the circumstances under which individualists are *not* able to enforce a modification of the structure. We have not examined

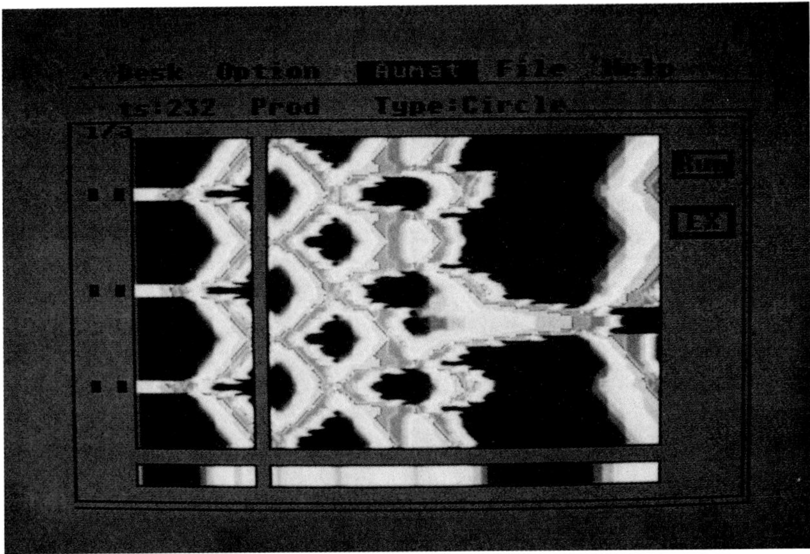

Figure 10: The progressive behaviour of the individualists leads to a chaotic development of the automaton

Figure 11: The regressive behaviour of the individualists leads to a low level of productivity

the question regarding how far the behaviour of individualists and their ability to enforce a modification are changed by an exchange of information, or by a real network of individualists. All these questions are of great interest, also in fundamental terms, and perhaps it will be possible to find answers in future with the help of a more sophisticated version of the described model. *Of this we are sure !*

References

[WEH90] S.Wehowsky: GEO-WISSEN, Chaos und Kreativität, Heft 2 (1990) 152

[BAC90] K.Bachmann: GEO-WISSEN, Chaos und Kreativität, Heft 2 (1990) 88

[BAK91] P.Bak, K.Chen: Spektrum der Wissenschaft, Heft 3 (1991) 62

[SIL88] G.Silverberg, G.Dosi, L.Orsenigo: The Economic Journal **98** (1988) 1032

[BAU89] W.J.Baumol, J.Benhabib: J.Economic Perspectives **3**, 1 (1989) 77

[EBE91] W.Ebeling: Syst.Anal.Model.Simul. **8**, 1 (1991) 3

[CAL89] V.Calenbuhr, J.L.Deneubourg: Actes coll.Insects Sociaux, **5** (1989) 207

[KUE84] R.Kühne: Physik in unserer Zeit (1984) 84

[LUH87] N.Luhmann: Soziale Systeme, Grundriß einer allgemeinen Theorie, Suhrkamp Taschenbuch Wissenschaft StW 666, Suhrkamp Verlag, Frankfurt am Main (1987)

[JIM80] M.A.Jimenez Montaño, W.Ebeling: Collective Phenomena **3** (1980) 107

[WOL84] St.Wolfram: Physica **10D** (1984) 1

[TOF87] T.Toffoli, N.Margolus: Cellular Automata Machines, The MIT Press, Cambridge Mass., London Engl.

Improbable Events in Deterministically Growing Patterns

P.J. Plath and J. Schwietering
Universität Bremen, Bremen

1 Introduction

The pigmentation pattern of shells of mono and bivalved molluscs are really fascinating to look at. Several attempts have been made to model these patterns [WAD69, LIN82, MEI84, MEI87, MEI91]. The activator- substrate or the activator-inhibitor models of H. Meinhardt [MEI87, GIE72, MEI82], which are based on coupled one-dimensional differential equations with diffusion terms, describe especially very well the generation of these pigmentation patterns. Furthermore, using the differential equations, it would be very difficult to take into consideration the noise on the variables which is caused by the environment.

Shells grow at their outer lips. Growing simply means that a new generation of cells is created after a time interval has passed. To model this growth, H. Meinhardt used a continuous one-dimensional space. In this type of models, one-dimensional differential eqations have to be solved for each time slice.

Time and space are continuous variables in these differential equations and they have to be discretised in order to be solved numerically. This discretisation neither reflects the cellular character of the shells nor the stepwise creation of the new generation of cells.

It was a great challenge for us to model the pigmentation patterns of the seashells just taking into account the discrete nature of the propagation of the cells and to simulate the growing of the pattern almost as successfully as H. Meinhardt, who used a quite different mathematical ansatz.

2 The Model

The mathematical model for our simulation is a one-dimensional vector automaton. This means, we take

- a path graph P_l or a circular graph C_l with l vertices.

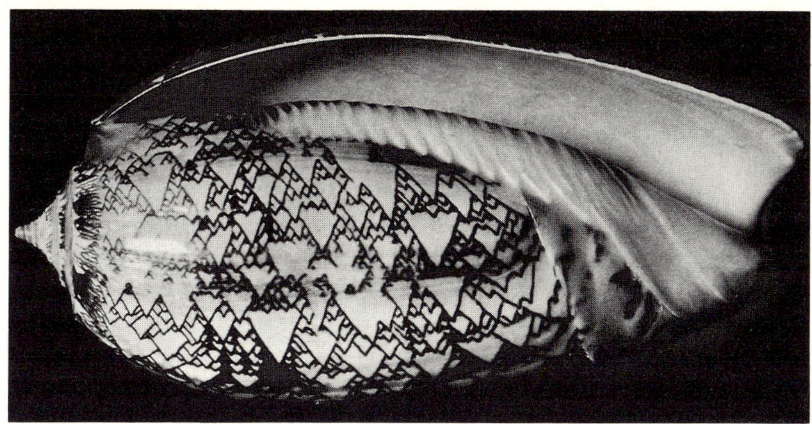

Figure 1: Oliva porphyria L.

- Each vertex is labelled by a vector in a two-dimensional concentration/ phase space, and is called a cell; The two components of the vector of the i-th cell at time $t \in \mathbb{N}$ are: the concentration $x(i,t) \in \mathbb{N}$ and the phase $p(i,t) \in \{0,1\}$.
- With respect to the temporal development of the cells a local neighbourhood is defined consisting of the actual cell and its adjacent vertices or cells repectively.
- Various transformation rules can be formulated which will transform both components of the vector from t to $(t+1)$: $x(i,t) \rightarrow x(i,t+1)$ and $p(i,t) \rightarrow p(i,t+1)$, depending upon the neighbouring cells.

These formal aspects define a one-dimensional cellular vector automaton. Let us now translate these ideas into a model of the biological system under consideration.

We assume that by propagation the creation of a cell in the new front of the outer lip of the shell at time $(t+1)$ depends on the current situation of the mother cell i in the actual front at time t and the actual situation of the cells at time t neighbouring the mother cell. the temporal sequence of states of the one-dimensional automaton thus represents the development of the moving front in the growing shell.

We assume that the incorporation of pigments in a new cell i at time $t+1$ (daughter cell) which is born from its mother cell i by cell division depends upon the concentration $x(i,t)$ of the reactants (prepigments) in its mother and in both of her neighbouring cells $i-1$ and $i+1$, which are the aunt cells to the daughter cell.

Moreover, it is reasonable to assume that each cell i can exist in at least two different states of activity, phases $p(i,t) \in \{0,1\}$ at time t. The activity of the mother cell will influence the amount of prepigments $x(i,t+1)$ which the daughter cell inherits from her mother. On the other hand, the daughter's activity $p(i,t+1)$ at time $t+1$ is determined by the concentration of the prepigments of her mother $x(i,t)$ and her aunts $x(i-1,t)$ and $x(i+1,t)$ as well as by the activity $p(i,t)$ of her mother.

Therefore the situation $z(i,t)$ of a cell i at time t is characterized by a vector

$$\vec{z} = z(i,t) = (i,t,p,x) = \begin{pmatrix} x(i,t) \\ p(i,t) \end{pmatrix} \tag{1}$$

We can formulate a transformation rule T, which determines the situation $z(i,t+1)$ of the daughter cell i at time $t+1$.

$$T : z(i,t) \mapsto z(i,t+1) \tag{2}$$

$$z(i,t+1) = \begin{pmatrix} x(i,t+1) \\ p(i,t+1) \end{pmatrix} \tag{3}$$

$$= \begin{pmatrix} f(p(i,t), x(i-1,t), x(i,t), x(i+1,t)) \\ g(p(i,t), x(i-1,t), x(i,t), x(i+1,t)) \end{pmatrix} \tag{4}$$

f and g are discrete functions which can be represented by $2 * k$ matrices:

$$\begin{pmatrix} f(0,0) & f(0,1) & f(0,2) & \cdots & f(0,k) \\ f(1,0) & f(1,1) & f(1,2) & \cdots & f(1,k) \end{pmatrix}$$

with $f(p,m) \in \mathbf{X}$; $m = 0,1,2,\cdots,k$; $p \in \{0,1\}$, where $\mathbf{X}$ is the set of the possible numbers (concentrations) of the prepigments $\mathbf{X} = \{0,1,2,\cdots\}$ and k is the largest number of prepigments which can be reached by the addition of the number of prepigments of the cell i and its neighbouring cells $i-1$ and $i+1$ at time t. The function g is given by:

$$\begin{pmatrix} g(0,0) & g(0,1) & g(0,2) & \cdots & g(0,k) \\ g(1,0) & g(1,1) & g(1,2) & \cdots & g(1,k) \end{pmatrix} \tag{5}$$

with $g(p,m) \in \mathbf{P}$, where $\mathbf{P}$ is the set of possible activities of a cell: $\mathbf{P} = \{0,1\}$. This transformation T is performed at the same time t for all cells of the automaton.

To obtain a temporal pattern one has to introduce some special cells into the starting automaton at time $t=0$ whose situation differs from that of all others. At least one cell j should have a small number of prepigments $x(j,0) > 0$, while the number of prepigments should be zero in all the other cells. However, all cells i of the automaton may have the same high activity $p(i,0) = 0$.

For colouring, the prepigments have to be transformed into the pigments. This process may depend upon the activity and the number of prepigments in the cell. Even the same pattern of prepigments can be coloured differently, accentuating special amounts of prepigments or the activity of the cells, or only

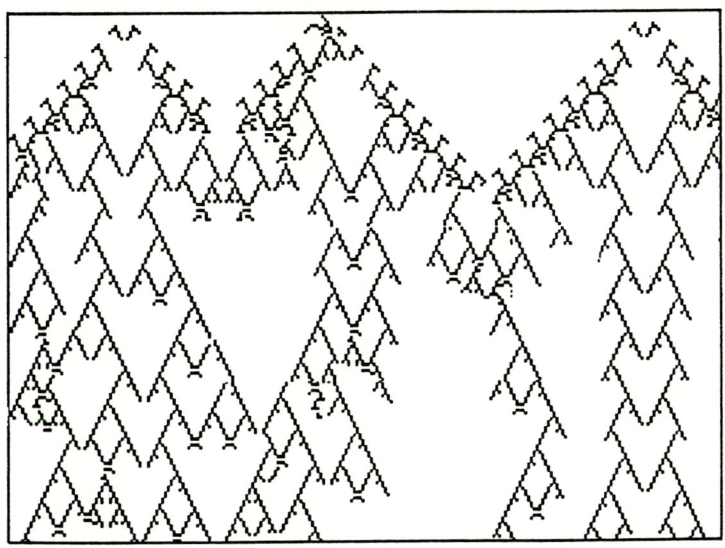

Figure 2: Purely deterministic pattern created by three exposed cells with $x(i_1, 0) = 1$, $x(i_2, 0) = 2$ and $x(i_3) = 1$. m is simply the sum of the values $x(i-1, t)$, $x(i, t)$, and $x(i+1, t)$ and $c(i, t) \neq 0$, if $x(i, t) = 4$ (black); the boundary conditions are: $x(l+1, t) = 0$ and $x(0, t) = 0$.

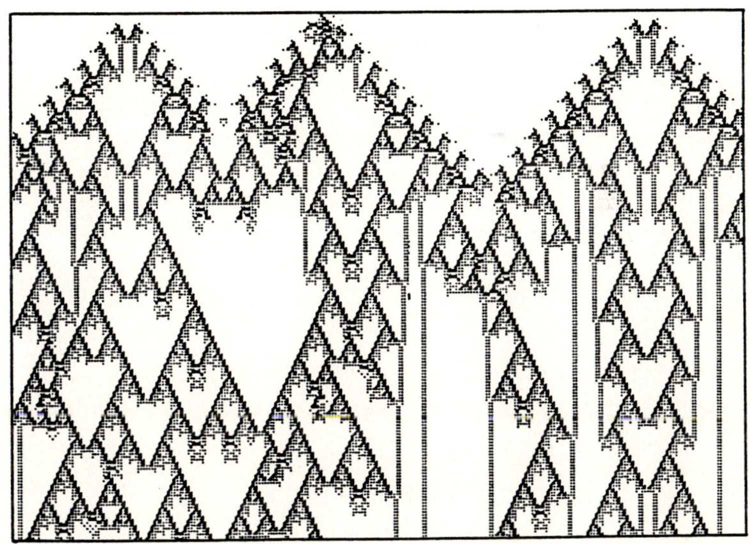

Figure 3: Purely deterministic pattern created by three exposed cells with $x(i_1, 0) = 1$, $x(i_2, 0) = 2$ and $x(i_3) = 1$. m is simply the sum of the values $x(i-1, t)$, $x(i, t)$, and $x(i+1, t)$ and $c(i, t) \neq 0$, if $x(i, t) = 4$ (black) or $x(i, t) = 3$ (grey); the boundary conditions are: $x(l+1, t) = 0$ and $x(0, t) = 0$.

a different way of transforming the situation of a cell into its pigment colour. Therefore, one has to define another discrete function which maps the components $x(i,t+1)$ onto a colour $C : x(i,t+1) \mapsto c(i,t+1)$. In this way a large variety of coloured patterns can be produced, some of which may resemble the observed seashell pattern.

The automaton is started by setting randomly the components $x(i,0)$ of a very few cells to be not equal to zero: $x(i,0) \neq 0$, whereas the components $x(j,0)$ are equal to zero for all other cells $j \neq i$. In any case, the phases of all cells are $p(i,0) = 0$ when the automaton is initialised. From the arbitrarily chosen initialisation the automaton starts and develops deterministically.

In order to create an improbable event, in each generation or time step we ask for a random number in the range between 0 and 99. If this number is less than the chosen threshold n, one of the cells of the t's generation will be selected randomly. The deterministic value $x(i,t+1)$ of the chosen i-th cell will randomly be increased by one. With this procedure, the natural fluctuations in the concentration of prepigments in a cell are reflected.

3 Results

Let us firstly consider the pigment pattern of the famous shell Oliva phorphyria L. (see Fig. 1). [ANG69, MEI91]. This shell is characterised by its fractal shape, which reminds one of solitary waves and Sierpinsky patterns. There are other shells, such as Aulicina vespertilio (see Fig. 4), which exhibit patterns much more similar to the Sierpinsky triangles than the pattern of Oliva porphyria.

Figure 4: Aulicina vespertilio, height 9.5 cm; the shell is characterised by its fractal pattern.

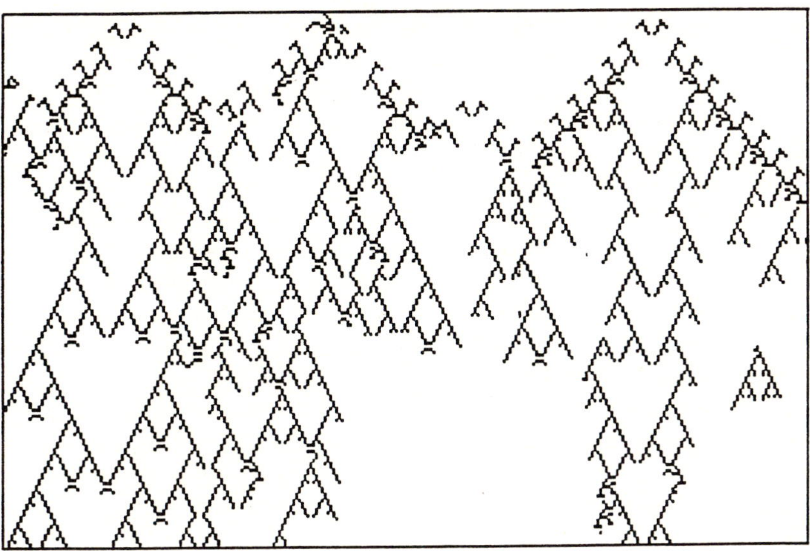

Figure 5: Probabilistic automaton initialised, governed and coloured by the rules of the automaton in Fig. 2, but with n=6.

Figure 6: Probabilistic automaton initialised, governed and coloured by the rules of the automaton in Fig. 3, but with n=6.

Figure 9: Cymbiolacca wisemani B. (1870); in this photograph the shell is is growing from bottom to top; the pattern shows white triangles, the basic side of which is coloured dark brown.

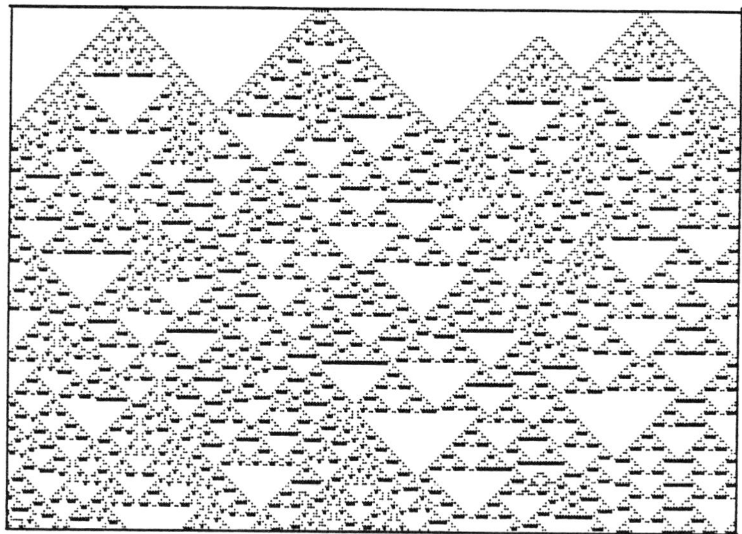

Figure 10: Simulation of the pattern of the shell Cymbiolacca wisemani B. (1870) by a stochastic cellular automaton with $x(i_1, 0) = 1$, $x(i_2, 0) = 2$, and $x(i_3) = 1$, and n=6. m is the sum of the values $x(i-1, t)$, $x(i, t)$, and $x(i+1, t)$, and the automaton is coloured by a special function; the boundary conditions are $x(l+1, t) = 0$ and $x(0, t) = 0$.

This pattern (see Fig. 9) [WIL71b] exhibits white Sierpinsky triangles, where the upper basic side consists of a few dark brown pieces of lines. The transformation rule for the simulating automaton of this shell pattern is very simple indeed, since the automaton works only in the active phase.

As usual, the pattern is very sensitive to the variation of this function. However, there are some positions in the coding of the transformation rule, which would produce very similar patterns. This example offers the possibility of finding an interpretation of the transformation rule of our one- dimensional cellular vector automata. The transformation rule might be understood as a genetic coding of the behaviour of the biological cells. The heading of the code has a very simple meaning: if we know the value of $x(j,t)$, where $j = i+1$ and $j = i-1$ for the neighbouring cells, than we know, what to do. The body of the code will state how to estimate the future state and phase. If there is a slide mutation in the code, in general a different pattern will result.

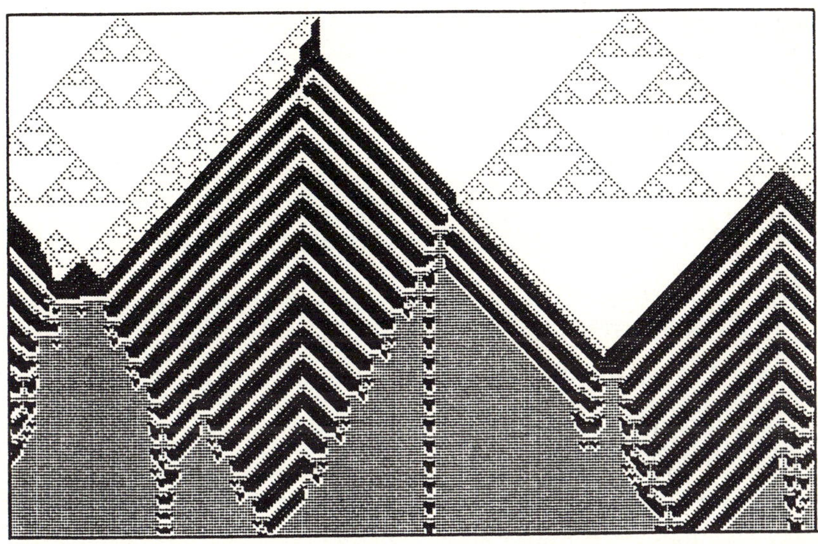

Figure 11: Probabilistic automaton created by three exposed cells with $x(i_1,0) = 1$, $x(i_2,0) = 2$, and $x(i_3) = 1$, with n=6. The transfomation rule consists of two parts connected by the improbable events. m is the sum of the values $x(i-1,t)$, $2 * x(i,t)$, and $x(i+1,t)$. The automaton is coloured by a special function; the circular boundary conditions are $x(l+1,t) = x(1,t)$ and $x(0,t) = x(l,t)$.

Let us finish with a very interesting example with respect to the interpretation of the transformation rules of the automata (see Fig.11). This special rule consists of two parts. Starting with the initialisation which has been used all along, only one part of the rule is firstly realised by the well-known growing Sierpinsky triangles. But because of the occurence of the improbable events the second part of the rule is switched on after some time and one can observe a quite different pattern.

The improbable events dicussed above are not mutations of the rules, but reflect the small fluctuations in the concentration of prepigments of the cells which might be caused by the environment.

5 Acknowledgement

The photograph in Fig. 1 is reproduced with the kindly permission of Grange Batelière - Paris. The photographs in Figures 7 and 9 are reproduced with the kindly permission of T.H.F. Publication INC. Hong Kong. We are therefore very much indebted to both publishing houses.

References

[WAD69] C.H. Waddington, J. Cowe: J. theor. Biol. **25** (1969) 219

[LIN82] D.T. Lindsay: Differentiation **2** (1982) 32

[MEI84] H. Meinhardt: J. Embryol. ex. Morph. **83** Suppl. (1984) 289

[MEI87] H. Meinhardt, M. Klinger: J. theor. Biol. **126** (1987) 63

[MEI91] H. Meinhardt, M. Klinger: Spektrum der Wiss. Heft 8 (1991) 60

[GIE72] A. Gierer, H. Meinhardt: Kybernetik **12** (1972) 30

[MEI82] H. Meinhardt: Models of biological pattern formation, Academic Press, London (1982)

[ANG69] S. Angeletti: Les Coquillages, Grande Batelière, Paris (1969) p.59, plate 102

[SWI91] J. Schwietering, P.J. Plath in: Modern Trends in Human Leukemia IX, Springer-Verlag Berlin Heidelberg New York London Paris Tokyo Hong Kong Barcelona Budapest (1991)

[KRI84] V.I. Krinsky: in Self-Organization Autowaves and Structures Far from Equilibrium, Springer-Verlag Berlin Heidelberg New York Tokyo (1984) (V.I. Krinsky editor)

[WIL71a] B.R. Wilson, K. Gillet: Australian Shells, Charles E. Tuttle Company, Rutland, Vermont & Tokyo, Japan (1971), p. 143, Plate 96.

[WIL71b] B.R. Wilson, K. Gillet: Australian Shells, Charles E. Tuttle Company, Rutland, Vermont & Tokyo, Japan (1971), p. 8, Plate 3.

Modeling Turbulent Gaseous Motion Using Time-Varying Fractals

Georgios Sakas

Technische Hochschule Darmstadt, Fachgebiet Graphisch-Interaktive Systeme

Abstract

This paper presents a new technique for modeling and animating turbulent gas motion using time-varying fractals. Our method works equally well in 2-D and 3-D and generates a stochastic, turbulent field that is variant over time and space. The proposed model is motivated by the stochastic spectral turbulence theory and employs spectral synthesis. Animation of gases is achieved through a phase shift in the frequency domain according to Kolmogorov's exponential law. The visualization techniques for both the 2-D and 3-D cases are described. Emphasis is placed on implementation on a multi-processor machine. An interactive, (quasi-)real-time, parallelized version, running within an X-windows environment, is described. Due to the fast feedback provided, this interactive version makes adjustment of the model parameters significantly easier.

1 Introduction

1.1 Motivation and Aim of the Work

Animation of turbulent gaseous movement remains a unsolved problem for computer graphics. Natural phenomena like rising smoke or steam and wind-driven clouds can be seen everywhere in nature. Thus, their visually appealing modeling and rendering is very important for visualizing high-quality outdoor animated scenes, or even for improving the performance of visual flying, ship and driving simulators. None of these "gaseous objects" can be handled with traditional polygon-oriented modeling and animation techniques because they cannot be approximated by polygons and, therefore, do not have anything like a surface, normal, etc. (although a surface can be virtually defined by considering the values of neighboring elements).

Given the importance of gaseous phenomena, we propose a new method for modeling and animating turbulent moving gases. The proposed model is motivated by the physics of turbulent motion and is in accordance with the research results achieved in that field. We employ the stochastic spectral synthesis method, which is applicable, without changes, in both the 2-D and 3-D cases and provides good visual results requiring only short computing time. Unlike the techniques proposed previously, a true animation of the fractal defining a turbulent field as a function of space and time is achieved. By means of the model the user can define different types of turbulent motion, different gases, velocities, etc. All parameters employed correspond to the physics-based properties of turbulent fields and are either already familiar to users or can be easily understood on the basis of everyday experience. Thus, all model parameters are intuitively comprehensible even to a user with a limited computer graphics background, so that the method is well suited for designers and animators. Most important, the user is able to predict the visual effect caused by the variation of a model parameter. Due to the computational effectiveness of the algorithm and its straightforward parallelization, a multi-processing implementation can be easily achieved. We present such a parallelized (quasi-)real-time version, running within an X-windows environment, with a graphical interface, which enables the interactive manipulation and adjustment of all model parameters and, thereby, an immediate feedback. In addition to its computational effectiveness, the method can be easily implemented in most modern rendering systems capable of handling solid textures.

1.2 Previous Works

1.2.1 Cloud Visualization

Voss [Voss85] modeled clouds using 2-D and 3-D fractals, while Gardner [Gard85] employed hollow ellipsoids and heuristic pseudo-random transparency texture functions. Willis [Will87] used Blinn's single-scattering approximation to improve the effects of daylight flying simulators. Max [Max86] used shadowing polyhedra to calculate the atmospheric scattering caused by light beams coming through gaps between clouds or leaves. Nishita et.al. [NiMN87] used Blinn's [Blin82] single-scattering model in combination with shadowing polyhedra and non-uniform densities. Klassen [Klas87] dealt with the visualization of the atmospheric effects. Sakas et. al. [Saka90], [SaKe91], [SaGe91] proposed a method for efficiently rendering arbitrarily distributed discrete volume-densities (voxel fields). Kajia/Herzen and Inakage [KaHe84], [Inak89] used ray-tracing to render volumes. All of these authors assumed "static" volumes and therefore did not address the problem of volume animation.

1.2.2 Cloud Animation

Kajia/Herzen presented a physics-based model for modeling and animating clouds. Their model is rather primitive and lacks details and optical complexity. Yaeger et.al. [YaUM86] used a semi-empirical model, based on simplified physics and measured data, to define a turbulent velocity field for driving several million particles along different stream lines (particle-systems approach). This approach requires a huge amount of preparation and manual input, enormous computation times (CRAY-II), and postulates the availability of valid, measured data. Such data are available only in limited cases. Even then, they are not always suitable for modeling purposes.

In the functional approach turbulence is defined as a continuous function over space $turbulence = t(\mathbf{x}) = \sum | (1/r^{i}\, noise(r^{i}\mathbf{x})) |,\ r > 1$. Thus, the generation of such a function is achieved by a summation and overlapping of several, appropriately scaled down copies of a basic "stochastic primitive" grid, or noise function, called the integer lattice. The turbulence function can be defined in 2-D or 3-D space, so that bodies of arbitrary shape can be "sculptured out" of the texture (solid texturing). Perlin [Perl85] first introduced a functional synthetic turbulence model to assign different texture values to a reference body. Appealing animations of the solar corona have been achieved by translating and rotating the function definition space relative to the reference body, as well as through the use of look-up table operations. Perlin and Hoffert [PeHo89], as well as Saupe [Saup88], [Saup89], extended this method to 3-D space. Saupe, in particular, extended the rather heuristic formulas of Perlin to a true fractal model; in addition, he proposed several methods for turbulent animation of fractal clouds. Ebert and Parent [EbPa90] used a variant of Perlin's method for their realistic fog animation in "Going Into Arts".

In order to produce animated sequences of 2-D or 3-D fractal clouds, the above mentioned authors first define a fractal domain with 3 or 4 dimensions, respectively. The first 2 (or 3) dimensions of the fractal are interpreted as spatial dimensions, the last one as the time axis. Thus, during animation one "cuts slices" out of the fractal at the time-axis location corresponding to the time requested. As an extension, several authors (e.g., [EbPa90] and [Saup89]) perturb the plane of the slice, using a fractal interpolation instead of the linear one along the time axis, to achieve a more turbulent appearance.

The main advantage in using the functional approach rather than spectral synthesis, which is presented in the next chapter, is that a function can be evaluated only at the locations needed during rendering, thus saving memory space. In addition, the "radius" of the texture evaluation can be adapted to the picture resolution and the distance of the object from the eye-point, thus anti-aliasing the texture during its generation (see [SaGe91]). On the other hand, the employed heuristic turbulence function is heuristic and requires parameters which cannot be intuitively understood by the user. The adjustment of these parameters in order to model a desired turbulence effect requires user experience and several trial-and-error runs. The main drawback of the functional approach lies in the turbulence animation itself: one dimension of the static data field is interpreted as the time axis, and movement is achieved by moving the points of the reference body along different paths through the static field. Although the visual results can sometimes be appealing, this still remains a heuristic and inflexible approach emulating rather than simulating true 3-D turbulent motion. In addition, there are only two parameters influencing the structure of the fractal field, namely fractal dimension and lacunarity. Thus, the manipulation possibilities are rather limited.

2 Modeling Techniques

2.1 Stochastic Spectral Synthesis

In contrast to Euclidean-space methods, spectral synthesis defines texture in *Fourier space* or *frequency domain.* Euclidean and frequency representations of a (periodic) texture function are coupled by means of the Fourier transformation FT

$$f(x,y) = \sum_{u=-N/2}^{N/2-1} \sum_{v=-N/2}^{N/2-1} F(u,v)\, e^{i2\pi(ux+vy)/N}, \quad x, y = 0, 1, \ldots, N-1 \tag{1}$$

$$F(u,v) = \frac{1}{N^2} \sum_{x=0}^{N-1} \sum_{y=0}^{N-1} f(x,y)\, e^{-i2\pi(ux+vy)/N}, \quad u, v = -\frac{N}{2}, \ldots, \frac{N}{2}-1$$

The two representations of the discrete (texture) function are equivalent. This means that it is possible to transform a given function into frequency space, or to define a function in the frequency domain and obtain the Euclidean-space function by means of inverse transformation [GoWi87], [Brac65].

According to the notation of eq. (1), the Fourier transformation of a real function is represented by a series of complex factors (coefficients) $a + ib$, or in polar coordinates:

$$a + ib = r\, e^{i\phi}, \qquad r = \sqrt{a^2 + b^2}, \qquad \phi = arctan\,(\frac{b}{a}) \tag{2}$$

The magnitude r of a Fourier coefficient is the amplitude of the corresponding term (wave) in the Fourier transform, whereby the phase angle ϕ determines the shift of the wave with respect to the origin of the coordinate system. The Euclidean distance of a spectrum coefficient from the origin is called its frequency. In the case of a time-varying signal the frequency f has units of cycles per unit time, while in the case of a space-varying signal the frequency k is measured in cycles per unit length. A stochastic spectrum is characterized by the form of its mean value as a function of the frequency and by the distribution of the (amplitude and phase) coefficients around this mean. If the sample (or the corresponding spectrum) is quadratic with a resolution which is a power of 2, the computationally much more effective Fast Fourier Transformation (FFT) can be employed.

Some properties of the Fourier transformation, which are important for the work presented here, are summarized in the following section. For a *real function* the following is valid:

$$F(u) = F^*(-u) \tag{3}$$

whereby $F^*(-u)$ is the conjugate complex value of $F(u)$. This property means that for real functions only one half of the spectrum has to be defined, while the other half is completed in accordance with eq. (3) (see also Figure 1). We use this property for simplifying the calculations: we transform only one half of the spectrum using FFT. The Fourier transform and its inverse are periodic:

$$F(u+N) = F(u) \tag{4}$$

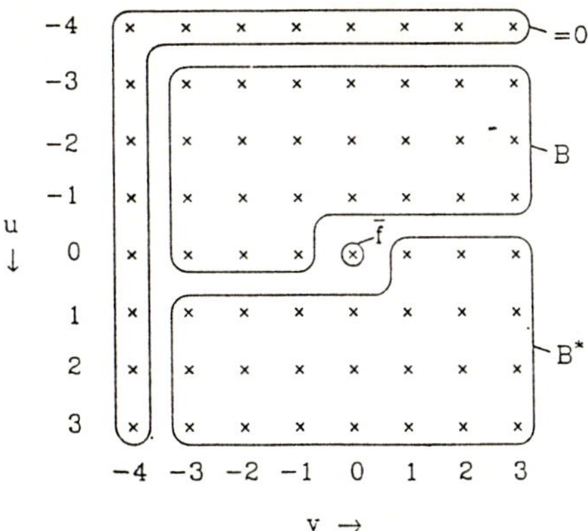

Fig. 1: Spectrum of a 2-D real discrete function showing the independent (B) and conjugate complex (B^*) parts; $\bar{f}$ is the average value

whereby N is the resolution. A translation in Euclidean space results in a phase shift in Fourier space, and vice versa:

$$f(x - x_0) = F(u)\, e^{-2\pi i u x_0 / N} \tag{5}$$

The average value of the function is given by F(0):

$$\bar{f} = \frac{1}{N} \sum_{x=0}^{N-1} f(x) = F(0) \tag{6}$$

and the mean square value:

$$\overline{f^2} = \frac{1}{N} \sum_{x=0}^{N-1} f^2(x) = \sum_{u=0}^{N-1} |\, F(u)\,|^2 \tag{7}$$

2.2 Basics of the Turbulence Theory

The statistical theory of homogeneous locally isotropic turbulence employed in this work was originally presented in the classic works of Reynolds, Kolmogorov, Obukhov, etc. An extensive body of literature exists for this difficult and interesting field. We recommend [Panc71] as an excellent, easy-to-read textbook and [FrMo77] and [TeLu72] for deeper analysis.

According to Reynolds, the velocity of a turbulent stream is regarded as the superposition of two motions: $u = \bar{U} + u'$, whereby $\bar{U}$ is an average translative velocity and u' an additional random fluctuating motion. The fluctuating motion is regarded as a result of the simultaneous existence of turbulent perturbations, or "eddies", of different sizes λ, with each eddy having a characteristic velocity of u_λ. Together, all of these eddies build a cascade along which energy is transformed from the basic to the turbulent motion, after which it diffuses to heat due to friction. According to Kolmogorov's hypothesis, the influence of friction in a turbulent current is negligible for eddies of all scales with the exception of the very smallest. That means that energy is not generated or lost, but only redistributed among the various wave numbers along the cascade. The range of scales, in which this hypothesis is valid is called the *inertial subrange*.

When the spectral method is employed for the study of turbulence, the energy density spectra for the velocity, pressure, temperature, etc. of a turbulent current can be calculated. Turbulence is a 4-dimensional phenomenon, involving 3-D spatial as well as temporal variation of structures. Therefore, one can distinguish between, on the one hand, frequency spectra, which describe the variation of the field structure over time and expressed as functions of f or $\omega=2\pi f$, and, on the other hand, wave-number spectra, which are expressed as functions of k and describe the spatial structure of a "snapshot" of a turbulent field. As pointed out by Lovejoy and Mandelbrot in [LoMa85], the second spectrum describing the structure of a static frame follows a $1/f^{\beta}$ distribution for the amplitudes and a random $[0, 2\pi)$ distribution for the phases (white noise). The problem now is to find a spectrum for the velocities. As a first approximation we applied here the "classic" 3-dimensional wave-number spectrum for the velocities formulated by Kolmogorov and Obukhov for homogeneous, isotropic, non-intermitting turbulence with a high Reynolds number. If $\underline{U} \gg u'$, Taylor's "frozen turbulence" hypothesis is valid and the frequency spectrum is regarded to be equal to the wave-number spectrum. Different expressions in better agreement with experimental data have been given by Yaglom, Kármán, Goltsin, Ogura, Pao, Heisenberg, Batchelor, etc. and can be found in [Panc71], pp. 186 - 246. A common characteristic of all spectra is that they include a wide inertial subrange with an exponential frequency dependency of the type $S(f) \sim 1/f^{\kappa}$. The general form of the spectrum is presented in Figure 2.

$$S_u(f) = 1.22\, \varepsilon^{\frac{2}{3}} f^{-\frac{5}{3}} \quad \Rightarrow \quad S_u(f) \sim u_{L_0} f^{-\frac{5}{3}} \tag{8}$$

2.3 What Is An Eddy?

As mentioned in the description of turbulence, above, an eddy is a local disturbance of a certain size and velocity within the velocity field. The term "local disturbance" means that an eddy is associated with a spatial location within the turbulent current, even if its accurate position is not given or is not of interest. An eddy can be visualized by a vortex, although eddies do not necessarily show the same rotational motion. On the other hand, when spectral theory is used, an eddy is represented as the velocity component of the spectrum corresponding to the given eddy size. Thus, every coefficient of a (discrete) spectrum is associated with an eddy of the same size (wavelength).

There is a significant difference between these two representations: A Fourier coefficient at a given wavelength is, by definition, an average which incorporates contributions from *all* eddies of the same size, independent of the location of the eddies within the examined turbulence domain. Thus, a Fourier coefficient has no sense of position in Euclidean space and, therefore, cannot express local spatial

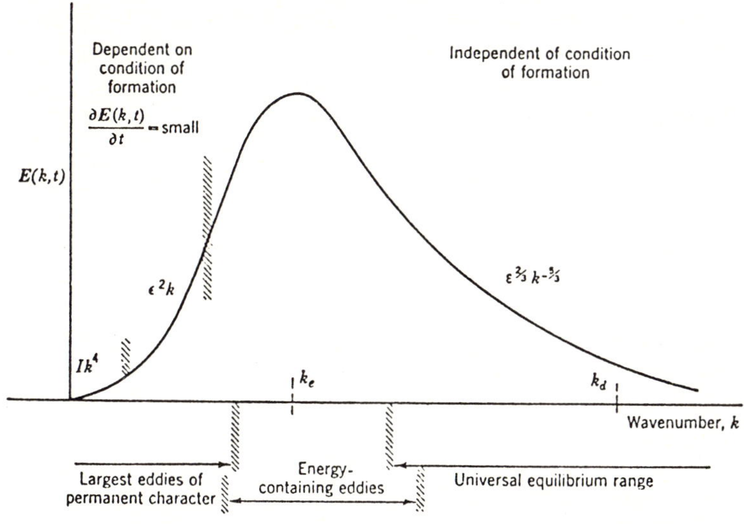

Fig. 2: General form of a turbulence spectrum (from [FrMo77])

structure. As a result, a Fourier coefficient should not be confused with a single eddy (see [FrMo77], pp. 133-146, for a more detailed discussion). Nevertheless, since we employ spectral synthesis we *do* refer to spectrum coefficients as eddies.

3 Computer Graphics Implementation

The implementation of the statistical theory of turbulence by the modeling and animation methods described above is rather simple. Due to the equality of Euclidean and spectral representations, spectral synthesis involves defining an adequate amplitude and phase spectrum and generating a desired texture by the inverse transformation. For this purpose, we generate random spectra with spatial and temporal distributions similar to those predicted by the spectral theory and confirmed by several experiments. Physics-based modeling and correct adjustment of the model parameters can be achieved by using one of the spectra given in the literature.

3.1 Static Frames

The reason for using spectral synthesis for defining (static) gaseous objects lies in the successful application of this method in the past for both analysis and generation of turbulent phenomena. As Mandelbrot and Lovejoy have shown ([LoMa85], [Mand75], etc.), turbulent fields are fractals with a fractal exponent $H \approx 0.7$. This result is in accordance with the statistical turbulence theory. An excellent overview about the use of fractals for describing and analyzing turbulence, including experimental results, can be found in [Sree91]. Concerning generative computer graphics, Voss ([Voss85], [PeSa88]) introduced the spectral synthesis of fractional Brownian motion in order to model extremely realistic static images of random fractal clouds.

3.2 Animation in the Frequency Domain

The most important reason for employing spectral synthesis is the advantageous and natural way of animating the static function. Since the function is expressed as the sum of all sinusoidal waves, or eddies, of the spectrum, an animation of the function can be achieved if each of the waves is animated. As can be seen in eq. (5), a translation (movement) in Euclidean space can be achieved over a phase shift in the frequency domain. The velocity u_f of each wave can be calculated by using the corresponding phase shift $\Delta\phi_f$ between two adjusted frames as follows:

$$u_f = \frac{\Delta\phi_f \, N}{2\pi f} = \frac{\Delta\phi_f \, N}{\omega} \tag{9}$$

Thus, the function is animated by shifting the phases of all waves for each new frame according to an adequate shifting law and then inversely transforming the result. This method has been used satisfactorily by Mastin et.al. [MaWM87] for animating sea waves. Since turbulence shows a strong stochastic character, a random shifting law is expected, as opposed to the deterministic law applying to the animation of sea waves. We use a velocity spectrum similar to Kolmogorov's (eq. 8), allowing thereby the original $\kappa = -5/3$ exponent to vary according to the wishes of the user. This procedure differs from the "slice method" used with the functional approach. The proposed method yields a soft shape transition of the gas and prevents jumps: large structures move in a rather translative manner and change slowly, continuously and smoothly, while small-scale structures randomize the visual appearance of the image. In addition, phase shifting involves simple addition and can be executed incrementally and very quickly.

3.3 The Model Parameters

For the implementation of the ideas described above we have developed a suitable computer graphics model containing seven intuitive parameters:

1) Fractal dimension H

2) Spatial structure size L

3) Translative velocity $\overline{U}$

4) Direction of translation $\vec{w}$ (wind direction)

5) Largest eddy size L_0

6) Velocity of the largest eddies u_{L_0}

7) Exponent of the velocity spectrum κ.

In the initialization step a random fractal field is constructed according to the fractal exponent H. By means of the spatial structure parameter L we can define the size of the largest spatial structure simply by setting all coefficients above this size to zero. The translative velocity $\underline{U}$ gives the overall speed of the current, while the vector $\vec{w}$ defines the translational direction. This velocity is applied to all waves. In addition, waves with frequencies lying between the largest L_0 and the smallest λ_0 eddies are perturbed according to the spectrum of eq. (8). A random velocity with a mean value $u(f) = u_{L_0}(f / f_{L_0})^{\kappa}$ is assigned to each of the waves with frequency f. The turbulence velocity u_{L_0} defines the speed of the biggest turbulent structure L_0.

Following the initialization of the spectrum and wave velocities as discussed above, velocities are calculated into phase shift values by means of eq. (9). In order to prevent aliasing, the value of a phase shift between two adjacent frames must be less than π. These phase shift values form the increment between two frames. As a result of eq. (3), for the synthesis of a real function, independent velocities have to be assigned only to one half of the waves; the other half is completed with the conjugate complex wave factors.

3.4 Visualization Methods

The method proposed above generates a stochastic field in 2-D or 3-D whose statistics obey certain laws characteristic of turbulent behavior. The elements of such a field are positive or negative real values, so that an additional transformation is necessary prior to rendering. For a 2-D or 3-D field one can map the texture values on the attributes of a body (surface or volume) by means of customary texture mapping techniques. Such attributes include color, transparency, brightness, density, etc.

3.4.1 2-D Visualization

The 2-D field is mapped onto polygons by using usual texture mapping methods. In order to visualize 2-D clouds we use the look-up table of our workstation to assign blue to the minimum and white to the maximum value of the random field and to compute all intermediate colors using a linear interpolation. Unfortunately, since we deal with an animated stochastic field, the minimum and maximum values change unpredictably during the sequence. This results in abrupt transitions of the mean cloud density between adjacent frames. Therefore, we designed the generation process to return directly "the majority" of the values in the desired 0-255 interval in order to address the look-up table without further rescaling. First, we define the offset (or mean value) of the function by means of F(0,0) according to eq. (6). Large values result in a cloud of high mean density, small values in a thin layer. By changing this value during the animation process and recalculating the look-up table, we can simulate the effect of condensation or dissipation of clouds in real time.

The second important parameter for the optical appearance of the turbulent field is the variance of the stochastic function. The variance is a measure of the spread of the function around the mean value and gives the probability of the generated values lying within the displayable interval 0-255. Texture values lying outside the 0-255 interval are clipped, resulting in "islands" of maximum or minimum density. As a result, large variance generates images of gases concentrated around several clusters, or islands, with large empty spaces between them, whereas small variance makes the gas appear to be rather homogeneously distributed around the mean value, resulting thereby in low-contrast cloud layers (see Picture 1). By means of eq. (6) and (7) one can easily calculate that the variance of a spectrum with the form r/f^{β} is proportional to r^2: $\sigma^2 = VAR(f(x)) \sim r^2$. Thus, a contrast variation of the cloud can be easily achieved over a variation of r.

3.4.2 3-D Visualization

The turbulence values are assigned as a density 3-D solid texture to a reference body in order to visualize 3-D fields. The reference body only defines the space occupied by the turbulence. We employ the illumination model and the scan-line volume rendering technique presented in [Saka90] and [SaGe91] to visualize the density field. The methods proposed there include a fast and effective illumination model, a fast traversing of the field by means of DDA and anti-aliasing in constant time by using pyramidal volume representations of varying resolutions. In order to avoid unpleasant sharp boundaries at the edges of the volume object, we scale the texture densities from the center to the edge by using an exponential function. Thus, the density of the volume object decreases gradually from the middle to the border. The shape of the used function can be seen in Figure 3.

3.5 Implementation

We designed an interactive interface for our model in order to encourage users, such as designers and animators, to generate their "personal" turbulences. Experience has shown that adjustment of parameters is usually a nerve-wracking and time-consuming trial-and-error procedure, whereby the user experiments with dozens of parameters until he finds the correct ones. With the interactive interface the user gets immediate feedback on the influence of a certain parameter on the optical result.

In our implementation we employed the widely used, standardized X-windows environment with OSF/Motiv widgets for the buttons, sliders, etc. Further, since the output of the program can be diverted to any other graphics workstation within a LAN network, we usually run the program on a powerful host (like the IRIS 380), while the windows manager runs locally. All parameters mentioned above can be adjusted by means of buttons, sliders, menus, etc., and the output can be directed to the screen or into a file. The local graphics capabilities of the workstation can be easily inserted into the program. We currently support the X and GL

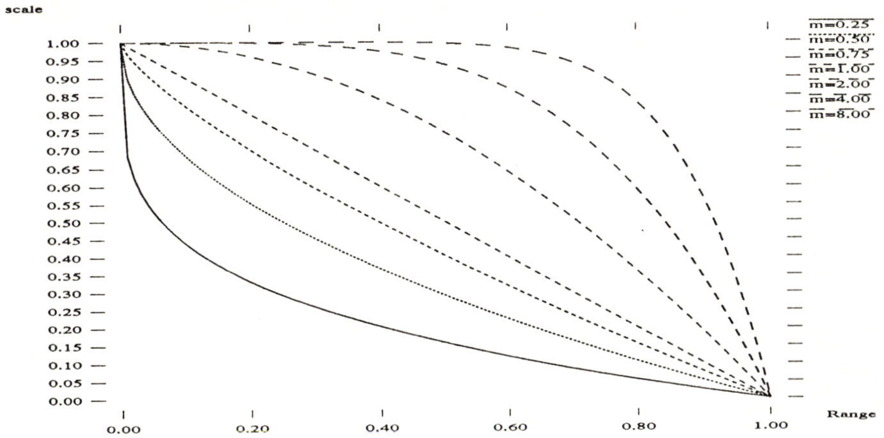

Fig. 3: Shape of the function used for scaling the 3-D texture data

libraries. Picture 6 shows this interface.

The complete cloud generation process consists of six steps:

1) Read new parameters for a cloud

2) Generate the spectrum of the static cloud

3) Calculate shifting tables

4) Move cloud using phase shifting

5) Create a real-number image using FFT^{-1}

6) Color and display the image

For our implementation we used an IRIS 380 VGX machine. This is a shared memory multi-processor MIMD computer with 8 CPU units and a powerful graphics engine. The architecture of the machine can be seen in Figure 4. A careful analysis of the sequential cloud generation process showed that over 85% of the computing time was required for calculating the inverse transformation, and for coloring and displaying the picture. Thus, we decided to parallelize this part of the computation.

We implemented heterogeneous parallelization for coupling the user interface with the calculation part, and homogeneous parallelization for the FFT transformation. The user-interface controlling process periodically reads the parameters adjusted by the user and registers all changes. This process is activated 20 times per second and, thus, requires a minimum of computing time leaving the processor free for other tasks. This results into an outer loop, recording parameter changes

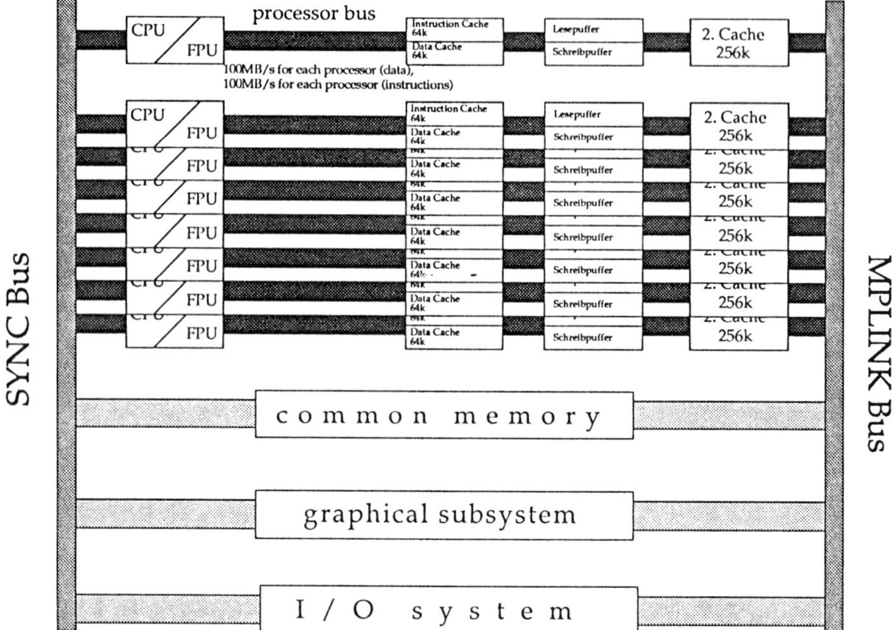

Fig. 4: Hardware architecture of the IRIS 380 VGX computer

and consisting from step 1, and an inner loop continuously calculating and displaying cloud images and consisting of steps 4 to 6. Steps 2 and 3 are initialization step activated only when one or more parameters change. If a change in any parameter is detected, the process first waits until the picture being calculated is finished. Then, the parameter value is updated and the appropriate action is taken. As an example, if wind velocity is changed, the new velocity will be considered only when the current picture has been completed. For any parameter change affecting the cloud movement, steps 3 to 6 of the pipeline mentioned above must be recalculated. If the seed of the pseudo-number generator or the fractal dimension H are changed, then the complete pipeline including step 2 has to be calculated again.

Homogeneous parallelization has been used for the FFT calculation. The 2-D FFT^{-1} transformation can be calculated by using a 1-D transformation, first over all rows, then over all columns; the 3-D transformation can be calculated similarly. Since each 1-D transformation may be calculated independent of all others, it is easiest to divide the N rows or columns of the spectrum into P blocks, with each block having N/P columns or rows. P is the number of the available processors; in our case P = 8. Since each processor has to calculate the same amount of data, we can expect that all processors will terminate approximately at the same time, so that a very simple process synchronization can be used. The parallelization of the

FFT can be seen on in Figure 5. After all rows have been calculated, the same scheme is used for calculating all columns. As an extension, we included the data coloration in the second loop. Thus, when the field transformation is completed, the data can be displayed directly. With this method we save an extra loop for coloring the cloud data.

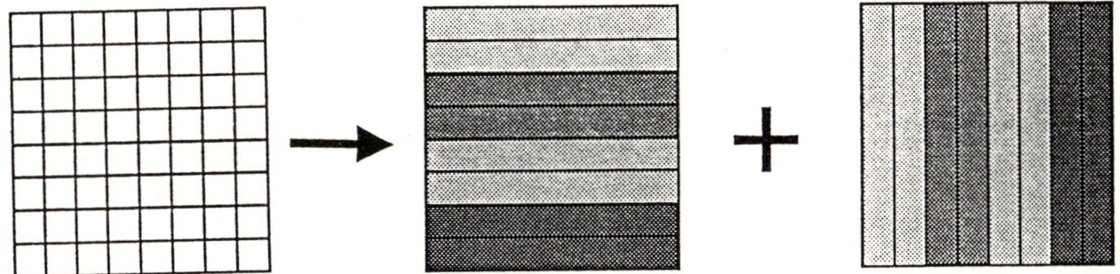

Fig. 5: Parallelization of the FFT^{-1} transformation

For displaying the transformed field we use the geometric engine of the machine. Since the IRIS architecture enables double-buffering, we switch buffers and immediately start with the computation of the next frame. Further, the VGX graphics engine enables Gouraud-interpolation between pixel values. Thus, we compute only 64^2 or 128^2 fields and use the hardware engine to interpolate up to ca. 1200 x 300 pixels on the display. According to our tests, a computation of 128^2 fields is sufficient for good optical results. In addition, we find it beneficial to "stretch" the quadratic field to fill a horizontally-oriented rectangular field, as seen in Picture 2. Such stretched fields appear much more realistic than quadratic ones. The reason for this is the nature of human perception: when we look to the horizon, clouds appear to be perspectively distorted. Therefore, animated clouds that are transformed to reflect that phenomenon look much more natural than non-transformed ones.

4 Results

The algorithm has been implemented on C and runs on any usual workstation, with or without specialized graphics hardware or multi-processing capabilities. Although our software is not yet optimal, the results are more than encouraging (please refer to Table 1). It is important to note that optimization or a different hardware will significantly reduce these times (please refer to the last section of this paper). With all 8 processors in action, real-time rates of about 20 pictures per second are achieved. In any case, the speed up is not linear, but roughly follows an exponential curve, as shown in Figure 6. The explanation for this phenomenon lies in Amdahl's law: increase in speed due to the use of P processors is

$$s \leq \frac{1}{F + \frac{(1-F)}{P}}$$

whereby F is the sequential portion of the process. The expected speed up is illustrated in Figure 6. The increase in speed, which we achieved by using 8 processors, yields an F of roughly 0.15, or a degree of parallelization of approximately 0.85. These results are in excellent agreement with our initial estimates.

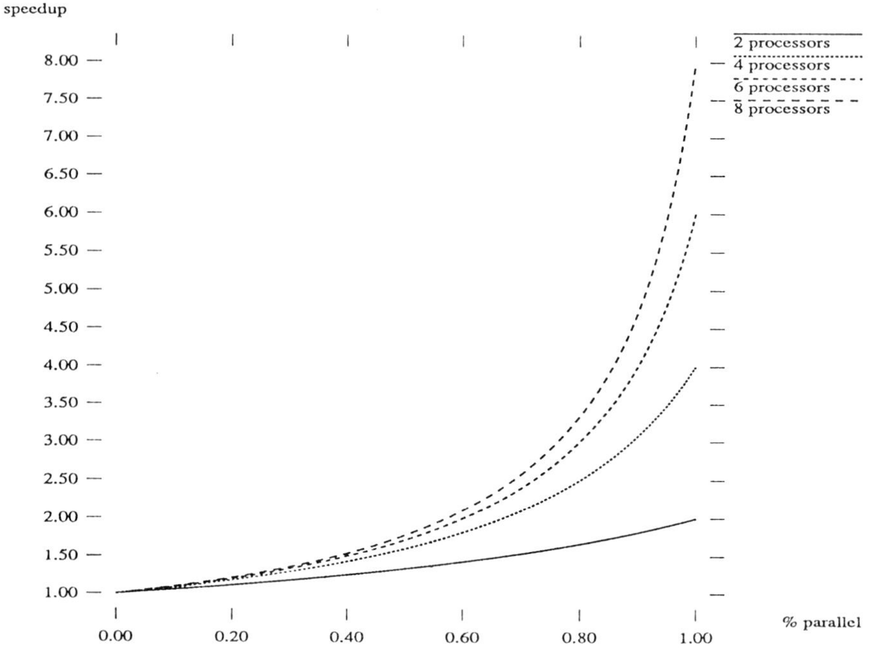

Fig. 6: Amdahl's law

Rendering and anti-aliasing the turbulent field works just as quickly when the methods presented in [Saka90] and [SaGe91] are employed. The bathroom scenes depicted in Picture 5 have been rendered using these methods at an image resolution of 720 × 576 pixels. For our general-purpose rendering system DESIRe, the typical times for rendering 128^3 voxel fields in TV-resolution lie between 3 to 10 minutes (including anti-aliasing). This is a significant improvement over the times achieved by ray-tracing.

5 Discussion and Further Works

This paper has presented a new method for modeling and animating turbulent gas motion which is based on the statistical theory of turbulence and employs spectral

Table 1: Runtimes for generating 2-D turbulent fields of varying resolutions with varying numbers of processors(in CPU seconds on the IRIS 380 VGX)

	Nr. of processors			
Resolution	1	2	4	8
32	0.05	0.03	0.03	0.02
64	0.18	0.1	0.07	0.06
128	0.85	0.5	0.3	0.25
256	4.5	2.5	1.6	1.15

Table 2: Speed up factors when using 2, 4 or 8 processors for generating 2-D turbulent fields of varying resolutions on the IRIS 380 VGX

	Nr. of processors		
Resolution	2	4	8
32	1.6	1.7	2.5
64	1.8	2.6	3.0
128	1.7	2.8	3.4
256	1.8	2.8	3.9

synthesis. Realistic animation of turbulent flow can be generated in 2-D or 3-D; in both cases a true time-dependent turbulence function is provided. The generated structures look "smooth": creasing is avoided and structures of any size appear to move smoothly and to change continuously. One has good control over the visual appearance of the texture: directional and oscillating properties, band limitation, "granularity" (= characteristic size of visual structure), etc. can be easily implemented. The chosen parameters increase design flexibility and even enable the employment of different spectra for the spatial and temporal turbulence structures. Since interaction with the model is facilitated by means of a small number of intuitive parameters, users with little experience in computer graphics - such as designers and animators - can utilize the method easily. The efficient model, combined with a fast evaluation algorithm written in C and running either on average or on super-workstations, includes an interactive graphics interface based on X-windows which makes the complete system convenient and, thus, attractive to users. A parallelization of the algorithm is straightforward and further increases the computation speed. Since fractal structures are considered to be the most effective method available for modeling natural phenomena, we felt free to apply familiar fractal techniques whenever possible, and to simplify, interpret and extrapolate the theory according to our needs wherever necessary.

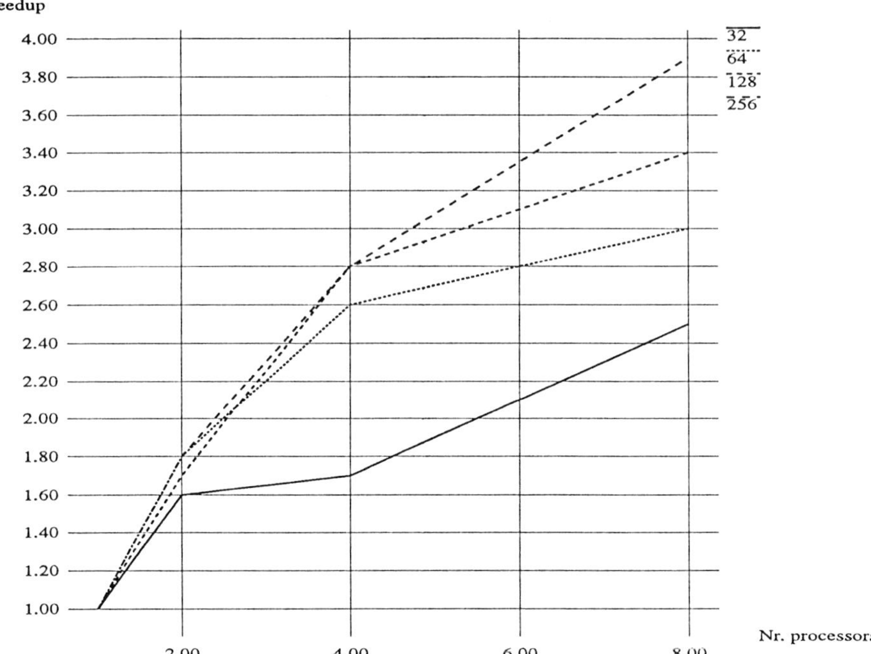

Fig. 7: Speed up factors when using 2, 4 or 8 processors for generating 2-D turbulent fields of varying resolutions

The shortcomings of our method are common to all spectral synthesis methods. Single values cannot be calculated unless the complete field is generated (*computational globality*). Another characteristic common to all texture-generating methods is that the generated texture does not - or does not naturally - interact with its environment: if a solid object (e.g., a hand or a lid) is placed within rising steam, the steam will flow *through* the object instead of around it. Although the scene modeler can partially avoid this effect by skillfully placing of the objects and the camera, or even by modeling a different path for the rising gas, this still remains a principal deficiency.

A serious limitation of the presented method is its *structural globality*. As pointed out in Section 2.2, the Fourier spectrum represents only averages of all local structures included in the turbulence domain; individual structures, like a single eddy, cannot be modeled. As a result, the texture is "uniformly defined", i.e., the optical characteristics are the same everywhere.

We are currently working on accelerating the computing times for field generation in order to approach real time for the 2-D case even on low-cost workstations. Because the vast majority of the computing time is spent on the inverse transformation, we are examining an inexpensive hardware support for accelerating this part of the calculation. Support of this sort can be provided by a fast digital signal processor already available on the market, by a modern CPU, like the i860, or by a specialized FFT chip. Results [1] gained with an optimized FFT version running on a single-processor i860 system are ca. 0.03 seconds for a 128^2 FFT^{-1}

transformation and ca. 0.5 seconds for a 512^2 field. In the 3-D case the improvements are even more impressive: a 64^3 field requires only 0.6 seconds. Thus, an i860 (or similar) solution seems to be far more effective and much cheaper than the general purpose workstation used. In addition, a second possibility is to increase the degree of parallelization of the program; this is currently about 85% and enables a maximum speed up of four when eight processors are used.

Acknowledgements

The author wishes to express his thanks to Prof. J. Encarnação and to all his colleagues and students for invaluable discussions and support during the implementation, to Dr. Krömker for continuous encouragement and tips, to Prof. Spurk for pointing out reference [Sree91], to Prof. Roesner for commenting on an early version of the manuscript, to J. Pöpsel for his friendly help (optimization of the FFT code), and to C. Distasio for proof reading. The parallel version of the algorithm has been developed by J. Dornauf during his diploma thesis, B. Kernke and J. Gosert tunned the system and helped by the development of the user interface; without their invaluable help this work had never been completed. This research has been carried out within the "Textur-Editor" project sponsored by the "Deutsche Forschungsgemeinschaft" (DFG; German Research Foundation), grant number EN 123/10-2.

6 Literature

[Blin82] Blinn, J. F.: *Light Reflection Functions for Simulation of Clouds and Dusty Surfaces*, ACM Computer Graphics, SIGGRAPH-82, Vol. 16, No. 3, pp. 21-29, 1982

[Brac65] Bracewell, R.: *The Fourier Transform and Its Applications*, McGraw-Hill, 1965

[EbPa90] Ebert, D., Parent, R.: *Rendering and Animation of Gaseous Phenomena by Combining Fast Volume and Scanline A-buffer Techniques*, ACM Computer Graphics, SIGGRAPH-90, Vol. 24, No. 4, pp. 357-367, August 1990

[FrMo77] Frost, W., Moulden, T.: *Handbook of Turbulence*, Volume 1: Fundamentals and Applications, Plenum Press, New-York, 1977

[Gard85] Gardner, G. Y.: *Visual Simulation of Clouds*, ACM Computer Graphics, SIGGRAPH-85, Vol. 19, No. 3, pp. 297-303, 1985

[1] All results for the i860 system have been provided by J. Pöpsel, AITEC corporation, Dortmund.

[GoWi87] Gonzalez, R. C., Wintz, P.: *Digital Image Processing*, Second Edition, Addison Wesley Publishing Company, 1987

[Inak89] Inakage, M.: *An Illumination Model for Atmospheric Environments*, R.A. Earnshaw, B. Wyvill (Eds), New Advances in Computer Graphics, Proceedings of Computer Graphics International, Springer-Verlag Tokyo, pp. 533-547, 1989

[KaHe84] Kajiya, J. T., von Herzen, B.: *Ray Tracing Volume Densities*, ACM Computer Graphics, SIGGRAPH-84, Vol. 18, No. 3, pp. 165-174, July 1984

[Klas87] Klassen, V.: *Modeling the Effect of the Atmosphere on Light*, ACM Transactions on Graphics, Vol. 6, No. 3, pp. 215-237, July 1987

[LoMa85] Lovejoy, S., Mandelbrot, B.: *Fractal Properties of Rain, and a Fractal Model*, Tellus, Vol. 37 A, No. 3, pp. 209-232, May 1985

[MaWM87]Mastin, G. A., Watterberg, P. A., Mareda, J. F.: *Fourier Synthesis of Ocean Scenes*, IEEE Computer Graphics and Applications, pp. 16-23, March 1987

[Mand75] Mandelbrot, B.B.: *On the geometry of homogeneous turbulence, with stress on the fractal dimension of the iso-surfaces of scalars*, Journal of Fluid Mechanics, Vol. 72, Part 2, pp. 401-416, 1975

[Max86] Max, N. L.: *Atmospheric Illumination and Shadows*, ACM Computer Graphics, SIGGRAPH-86, Vol. 20, No. 4, pp. 117-124, August 1986

[NiMN87] Nishita, T., Miyawaki, Y., Nakamae, E.: *A Shading Model for Atmospheric Scattering Considering Luminous Intensity Distribution of Light Sources*, ACM Computer Graphics, SIGGRAPH-87, Vol. 21, No. 4, pp. 303-310, July 1987

[Panc71] Panchev, S.: *Random Functions and Turbulence*, International Series of Monographs in Natural Philosophy, Vol. 32, Pergamon Press, 1971

[PeHo89] Perlin, K., Hoffert, E.: *Hypertexture*, ACM Computer Graphics, SIGGRAPH-89, Vol. 23, No. 3, pp. 253-262, July 1989

[PeSa88] Peitgen, H. O., Saupe, D.: *The Science of Fractal Images*, Springer Verlag, 1988

[Perl85] Perlin, K.: *An Image Synthesizer*, ACM Computer Graphics, SIGGRAPH-85, Vol. 19, No. 3, pp. 287-296, July 1985

[SaGe91] Sakas, G., Gerth, M.: *Sampling and Anti-Aliasing Discrete 3-D Volume Density Textures*, Proceedings EUROGRAPHICS'91, Vienna, Austria, pp. 87-102, North-Holland Publishers, September 1991

[SaKe91] Sakas, G., Kernke, B.: *Texture Shaping: A Method for Modeling Arbitrarily Shaped Volume Objects in Texture Space*, Proceedings 2nd EUROGRAPHICS Workshop on Rendering, Barcelona, 13-15 May 1991

[Saka90] Sakas, G.: *Fast Rendering of Arbitrarily Distributed Volume Densities*, Proceedings EUROGRAPHICS' 90, Montreux, Switzerland, pp. 519-530, North-Holland Publishers, September 1990

[Saup88] Saupe, D.: *Point Evaluation of Multi-Variable Random Fractals*, in: J. Jürgens, D. Saupe (eds.): Visualisierung in Mathematik und Naturwissenschaft, Bremer Computergraphik Tage 1988, Springer-Verlag, Heidelberg 1989

[Saup89] Saupe, D.: *Simulation und Animation von Wolken mit Fraktalen*, in: Informatik Fachberichte 222, M. Paul (Hrsg.), Proceedings GI-19. Jahrestagung I, Springer-Verlag, München, Oktober 1989

[Sree91] Sreenivasan, K.: *Fractals and Multifractals in Fluid Turbulence*, Annual Reviews of Fluid Mechanics, Vol. 23, pp. 539-600, 1991

[TeLu72] Tennekes, H., Lumley, J.: *A First Course In Turbulence*, The MIT Press, The Massachusetts Institute of Technology, 1972

[Voss85] Voss, R.: *Random Fractal Forgeries*, ACM Computer Graphics, SIGGRAPH-85 Conference Tutorial Notes, 1985

[Will87] Willis, P.: *Visual Simulation of Atmospheric Haze*, Computer Graphics Forum, No. 6, pp. 35-42, 1987

[YaUM86] Yaeger, L., Upson, C., Myers, R.: *Combining Physical and Visual Simulation - Creation of the Planet Jupiter for the Film "2010"*, ACM Computer Graphics, SIGGRAPH-86, Vol. 20, No. 4, pp. 85-93, August 1986

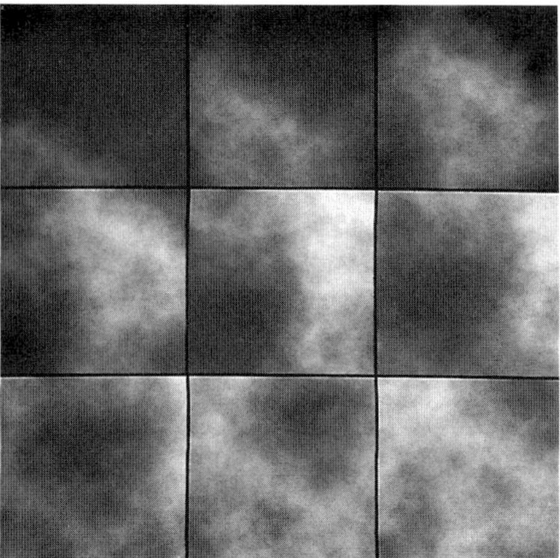

Pic. 1: Turbulent motion of 2-D clouds simulating condensation

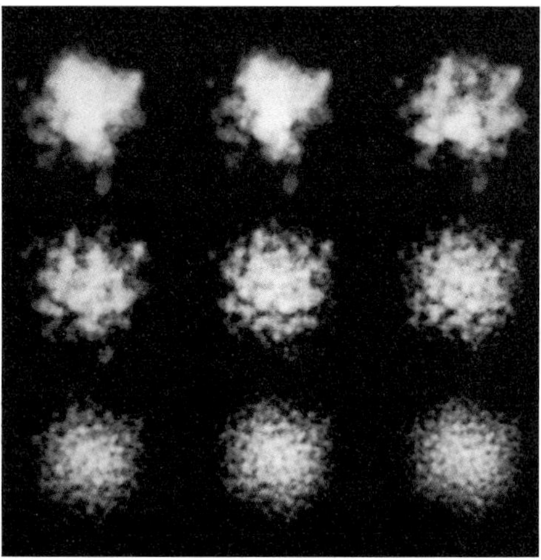

Pic. 2: Influence of the parameter L on 3-D clouds. The value of L increases from top left to bottom right

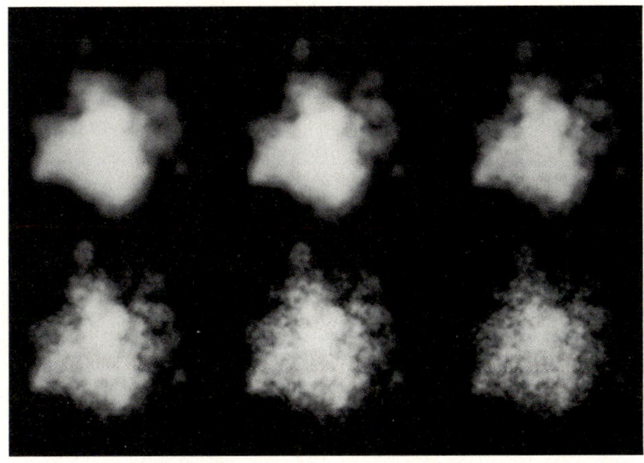

Pic. 3: Influence of the parameter H on 3-D clouds. The value of H increases from top left to bottom right

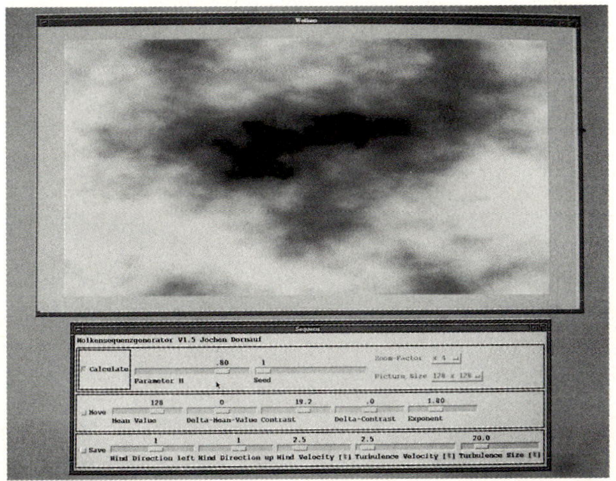

Pic. 4: User interface with real-time display

Pic. 5: Stills from the "foggy bathroom" sequence

Pic. 6: Stills from the "foggy bathroom" sequence

Devil's Gearworks

G. Mantica

Service de Physique Theorique, France

1. Abstract

Cylinders rolling on each other without friction, and at the same time filling the space between parallel planes show interesting fractal structures. These structures can be understood with the aid of suitable discrete subgroups of $SL(2, C)$, the group of Möbius transformations.

2. Introduction

Fractal models of natural phenomena are becoming more and more popular among physicists. One of these models, though, has a time honoured history: it deals with the problem of filling with circles the full area between two parallel lines, and is due to Apollonius of Perga [MAN]. As it turns out, Apollonius' solution (which we will describe later) is just a possible solution among the (infinitely) many which can be constructed [HMB]. In this paper, we will describe two particular families of such solutions, which fulfill an additional requirement: imposing a rotation on any one of the circles must set the whole structure in coherent (that is, slipless) rotation. In other words, the structure must behave like a perfect gearwork at all length scales. Hence, the well deserved name of *devil's gearworks.*

The physical relevance of these bizzarre geometries is widespread. In geophysical sciences, for instance, they can be used to model rocks between moving tectonic plates, whose fractal nature has been experimentally observed [SAM]. This description accounts for the lack of heat production in these regions, which can be attributed to the small friction induced by gliding friction [MCC].

The first geometrical fact needed to obtain a perfect gearwork is the slipless rotation of tangent circles. Consider an open chain of tangent circles, and let ω_j, r_j, $j = 1, \ldots, n$ be their angular frequencies and radii, respectively. It is easy to see that slipless rotation is equivalent to $\omega_1 r_1 = -\omega_2 r_2 = \ldots = (-1)^{n+1}\omega_n r_n$. If now we close the chain by imposing $\omega_1 = \omega_{n+1}$, $r_1 = r_{n+1}$, we see that a necessary and sufficient condition for slipless rotation is that *all possible loops of tangent circles must be composed of an even number of elements.* This requirement can be satisfied - for instance - buiding self-similar structures having "basic" four-element loops. We will now examine the explicit construction of these packings.

2. Geometrical Construction

The crucial mathematical tool we need for circle packings are Möbius transformations, that is, linear fractional transformations of the complex z-plane of the form $z' = (az+b)/(cz+d)$. The constants a, b, c and d will be written as the entries of a 2×2 matrix M, which fulfills $Det(M) = ad - bc = 1$. Möbius transformations are so important because they map circles into circles [BED]. They can be decomposed into translations, rotations, reflexions and inversions. In particular, an inversion about a circle Γ is a Möbius transformation that maps conformally the interior of Γ to the exterior, and vice versa, like images in cylindrical mirrors.

Let us now consider the basic four-circles loop, shown in Fig. 1a. It is composed of the circles A, B, and the straight lines C_0 and $\hat{C}_0$. This four circles are mutually tangent (one of the tangency point being at infinity). The packing is constructed mapping this basic loop *ad infinitum*, using the action of a discrete subgroup of $SL(2,C)$. To introduce the needed Möbius transformations, the geometrical parameters must be appropriately fixed. The largest circle B in Fig. 1a is tangent to the strip $\hat{C}_0$ at the point I_B, and to the circle A, which in turn is tangent to C_0 in I_A. R_A and R_B, the radii of A and B are then the only free parameters.

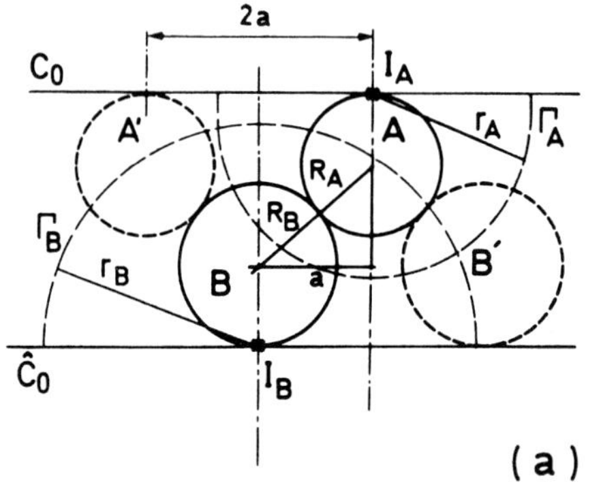

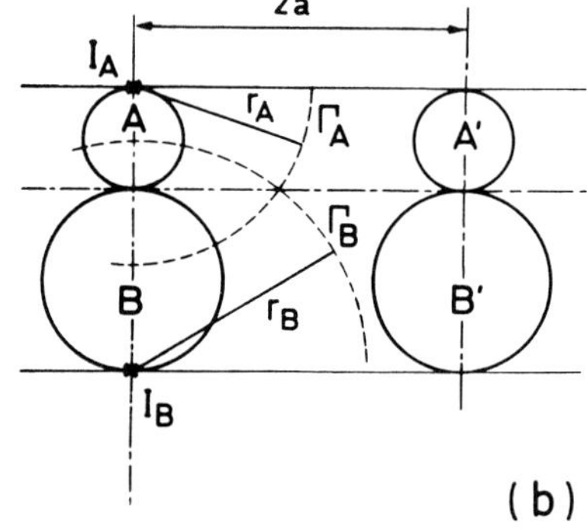

Fig. 1 The geometrical construction.

The first transformation we want to employ is (i) a translation $\mathcal{T}$ of step $2a$ parallel to the strip: $\mathcal{T}(z) = z + 2a$. To fix a, we impose that A be tangent to $\mathcal{T}(\mathcal{B})$, i.e.

$$2(R_A + R_B) = 1 + a^2. \tag{1}$$

This transformation iterates the "unit cell"' of the packing, moving toward infinity areas that are within the unit cell at the origin. To map such areas back into the first unit cell we employ two inversions: (ii) an inversion with respect to the circle Γ_B of center I_B and radius r_B, which leaves A invariant. We require that C_0 be mapped into B, which entails $r_B^2 = 2R_B$. It is mathematically convenient to left-multiply $\mathcal{I}_B$ by $\mathcal{R}$, a symmetry transformation with respect to the line $x = a$, and get a Möbius map whose matrix representation is

$$\mathcal{R}\mathcal{I}_B = \begin{bmatrix} 2a/r_B & -r_B \\ 1/r_B & 0 \end{bmatrix}.$$

(iii) The second inversion is effected with respect to the circle Γ_A of radius r_A centered at I_A. This inversion must leave the circle B invariant. We also require that $\hat{C}_0$ be mapped into A, and this leads to the equation $r_A^2 = 2R_A$. Since $(\mathcal{R}\mathcal{I}_A)^2$ is the identity, it cannot produce new circles. It is then convenient to take as third generator the composed map

$$\mathcal{R}\mathcal{I}_A\mathcal{T} = \begin{bmatrix} (a+i)/r_A & (a^2+1-r_A^2)/r_A \\ 1/r_A & (a-i)/r_A \end{bmatrix}.$$

We are still left eith the problem of fixing the free parameters in this construction. As it turns out, only very special (quantized) choices will produce a circle packing. Different choices, when applied on the basic circles A, B provide messy structures of overlapping circles. The conditions to be imposed to avoid this displeasing feature are that the Möbius transformations above must be translations, or discrete rotations with finite periodicity. In other words, their eigenvalues must be ± 1 or $e^{\pm 2i\pi/(n+3)}$. Since the matrices of our generators, $\mathcal{T}$, $\mathcal{R}\mathcal{I}_B$, and $\mathcal{R}\mathcal{I}_A\mathcal{T}$ have unit determinants, the above condition is equivalently stated in terms of the traces:

$$Tr\,\{\mathcal{R}\mathcal{I}_B\} = 2a/r_B = 2\cos(\pi/(m+3)), \tag{2}$$

and

$$Tr\,\{\mathcal{R}\mathcal{I}_A\mathcal{T}\} = 2a/r_A = 2\cos(\pi/(n+3)). \tag{3}$$

If $z_n = \cos^{-2}(\pi/(n+3))$, the previous relations and eq. (1) uniquely determine all the geometrical parameters as functions of the two integers m and n:

$$a^{-2} = z_n + z_m - 1; \;\; r_A^2 = a^2 z_n; \;\; r_B^2 = a^2 z_m. \tag{4}$$

In Fig. 2 we show a particular solution ($n = m = 3$). We notice that the integers m and n are counting the number of *necklace circles* tangent to the original circles A and $\hat{C}_0$ in the lower part of the strip, and tangent to C_0 and B in the upper part of the strip. It is also apparent that the rotation condition is fulfilled.

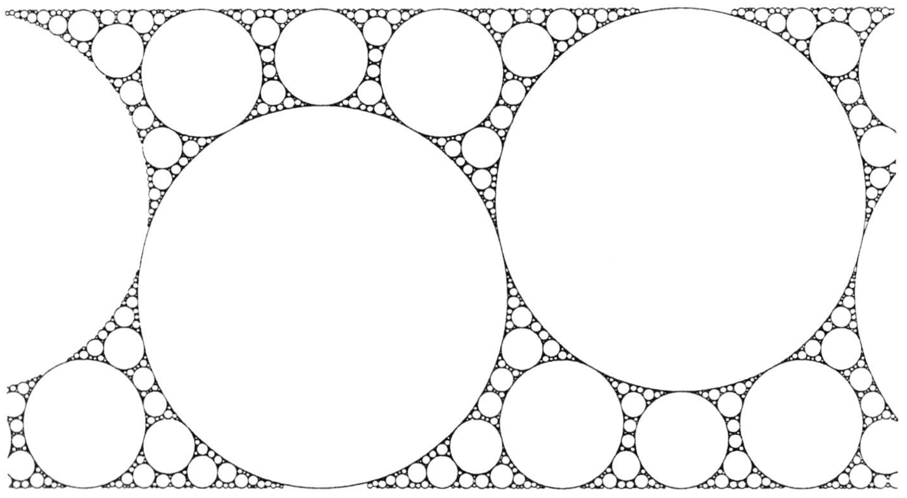

Fig. 2 The packing $n = m = 3$.

The advantages of this representation are many. They enable us to generate fractal packings in a deterministic, efficient fashion. For instance, we can positively know *all* circles whose radius is larger than any given threshold ε, a fact which is crucial for the computation of scaling indices, as we show in the following. Moreover, these models can be tuned to obtain different fractal dimensions, porosity, and so on, thanks to the infinite sets of solutions. Finally, they reduce to the classical Apollonian solution as n, m tend to infinity.

3. A second family

Also employing four loops, a second construction is possible. We choose a different set of basic circles, (Fig 1b) where now A is exactly placed on top of B. The generators $\mathcal{T}$ and $\mathcal{RI}_B$ are exactly the same as before, but $\mathcal{RI}_A\mathcal{T}$ is replaced by

$$\mathcal{RI}_A = \begin{bmatrix} (2a+i)/r_A & (2ia+1-r_A^2)/r_A \\ 1/r_A & -i/r_A \end{bmatrix}, \quad r_A^2 = 2R_A.$$

The trace conditions result again in eqs. 2 and 3, but the geometrical constraint (1) is now replaced by $2(R_A + R_B) = 1$, and (4) becomes: $a^{-2} = z_n + z_m$, $r_A^2 = a^2 z_n$, $r_B^2 = a^2 z_m$. By repeated application of the resulting transformations on the basic circles, one generates a complete filling of the strip. In Fig. 3, we show two packings belonging to this family. Only circles in the basic unit cell are shown.

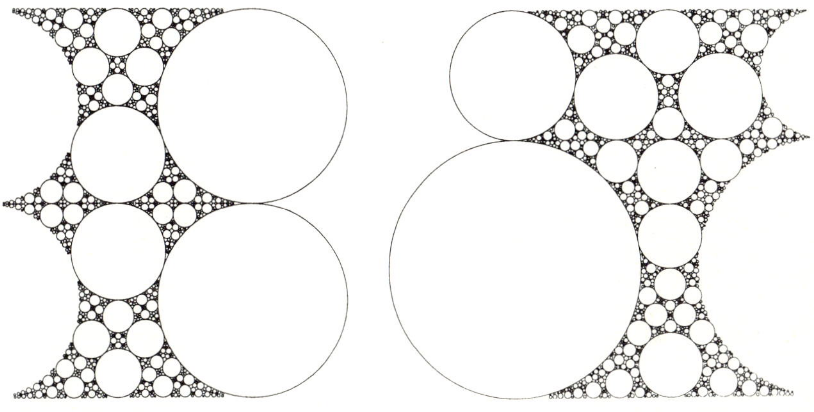

Fig. 3 Two packings of the second family.

4. Scaling indices and Conclusions

Our technique enables us to compute very efficiently scaling indices. We define the number, $N(\epsilon)$, of circles whose radius is larger than ϵ. We also let $s(\epsilon)$ be the sum of their perimeters ("surface"), and $p(\epsilon)$ be the porosity, that is, the area not covered by circles (of radius larger than ϵ), all per unit area. These quantities can be related to the fractal dimension d_f of the limit set of the structure, through

$$N(\epsilon) \sim \epsilon^{-d_f}, \quad s(\epsilon) \sim \epsilon^{1-d_f} \text{ and } p(\epsilon) \sim \epsilon^{2-d_f} \quad . \tag{5}$$

Typical values obtained for the fractal dimension in the first family increase with decreasing n and m and reach about 1.42 for $n = m = 0$. For the same labels, $n = m = 0$, the packing of the second family provides a value of about 1.52.

An interesting relation between this construction and turbulence can be obtained as follows. Since all discs rotate with the same tangential velocity v, the kinetic energy of discs of radius r is $E(r) \propto v^2 r^2 n(r)$, and the energy spectrum (as a function of the wavevector $k = r^{-1}$) is $E(k)dk \sim k^{\tilde{\tau}-4}dk = k^{d_f-3}dk$. Our model shows values of d_f ranging from 1.30 to 1.52, consistently with the Kolmogoroff scaling of the energy spectrum of homogeneous fully developed turbulence [BAT]: $E(k)dk \sim k^{-5/3}dk$. It will be interesting to investigate its relations with the β model of turbulence [FSN].

5. References

[BAT] G.K. Batchelor, *Theory of homogeneous turbulence* (Cambridge Univ. Press, 1982).

[BED] D. Bessis, and S. Demko, Comm. Math. Phys. **134** 293 (1990).

[FSN] U. Frisch, P.L. Sulem, M. Nelkin, J. Fluid Mech. **87**, 719-736, 1978.

[HMB] H.J. Herrmann, G. Mantica, and D. Bessis, Phys. Rev. Lett. **65** 3223 (1990).

[MAN] B.B. Mandelbrot, *The Fractal Geometry of Nature* (Freeman, San Francisco, 1982).

[MCC] W. McCann, S. Nishenko, L. Sykes and J. Krause, Pageoph **117**, 1082 (1979); C. Lomnitz, Bull. Seism. Soc. Am. **72**, 1441 (1982).

[SAM] C. Sammis and G. King and R. Biegel, Pageoph **125**, 777 (1987).

IV. Part: Picture Analysis

Dentronic Analysis of Pictures, Fractals and other Complex Structures

P. Hanusse, P. Guillataud

CNRS, Talence Cedex, France

1. Introduction

In this work we present a hierarchical geometric analysis, named *Dendronic Analysis*, designed to extract the semantic content of complex structures. It does not require any predefined model for the content of the "signal", thus implementing a totally data-driven self-structuring process.

After a few basic statements which describe the philosophy of the motivations, the goals and the context of this approach, we present a few concrete applications which examplify the concepts introduced below. These examples deal with pictures understanding, fractal curves and dynamical systems trajectories.

Let us first go straight into the conceptual core of this approach.

Modelling is telling something about our perception, writing something about it, in the proper sense "describing" it. If we are able to do it in a way that *respects* the essential nature of this perception, we produce a compact and relevant description, which depicts nothing else than the semantic information that "explains" the observation, that exhibits the objects it contains and their interrelationships.

Each object creates the space in which it lies as an object. Discovering this space amounts altogether to detecting the object and at the same time defining it. In that sense, on could speak of "self-definition". In most conventional modelling approaches, this self-definition process is not achieved because one most often impose the space in which we "force" the object to lie. To exhibit self-definition, the process of analysis must, from a neutral description, yet in some sense arbitrary, build up a space that is compatible with the internal structure of the observation. This can be performed only in a data-driven way. In this process semantic as well as symbolic information is extracted and at the same time data compression is achieved. This *numeric-symbolic* axis defines a space of hierarchical nature, in which levels of descriptions establish the various refinements of our perception. The very way in which the amount of information changes as a function of refinement requirements tells us something about the

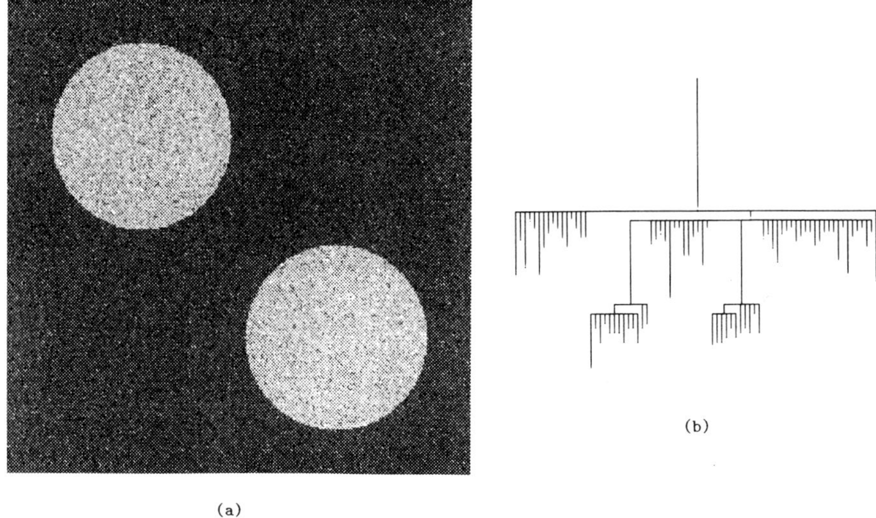

(a) (b)

Figure 1: A simple noisy picture (a) and its corresponding dendrone (b)

structure of the data and allows us to define the various levels of description of the observation.

We have named *Dendrone* or *Dendronic Structure* the hierarchical structure by which is coded the semantic information of an observation, of whatever nature, and *Dendronic Analysis* the general methodology, set of specific algorithms, that deals with the production and manipulation of this structure.

These very general and perhaps obscure statements will be made explicit and applied to various structures as already announced. In each case the formulation of such an approach has a specific form, which depends on the nature of the space in which the "signal" is initially given.

Let us consider a first example.

2. Detecting and defining objects in pictures

The understanding of perception is by definition the central problem of picture analysis. Much can be learned about modelling, in the general sense, from vision problems. Human vision has tremendous capabilities in this context. Making them explicit, as much as possible in the present state of our knowledge, may have an important impact in other fields where we so often try the discover the true nature of a signal [HAN90, HAN91].

Let us consider the picture in fig.1a. When asked to describe it, any observer will produce a statement such as *"two white disks on a black background"* and will perhaps add *"with some noise"*. This description is quite compact and obviously summarizes all that is significant about the picture. As compared to the amount of numeric information initially given in the digital picture, the data compression achieved in producing this description is quite dramatic, and, in a

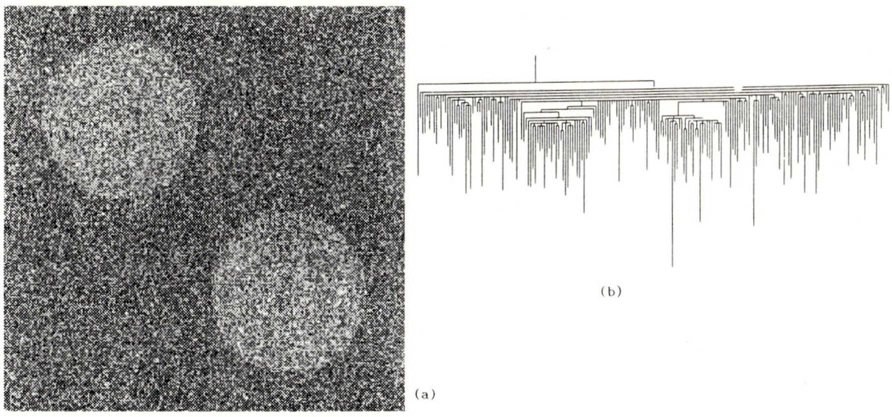

Figure 2: Same as fig.1 but with high amplitude noise

parallel fashion, the sense is correspondingly increased. Ideally we would like to design an algorithm that performs automatically this very task. In the same figure (1b) is depicted the dendrone that can be produced. Even before knowing how it is obtained and what it really means, one can see that it represents the very statement we are looking for: *there are two objects on a noisy background"*.

Let us explain how such a dendronic structure is produced.

In the present context, a digital picture is considered as a two dimensional light intensity field $z = I(x, y)$, as it is the most natural discrimination space that uses our vision system, in particular to analyse large scale objects. Note that the intensity field could also be considered as a two dimensional surface in three dimensional space, more appropriately noted $I(x, y, z) = 0$. Based on the idea that local contrast and connexity are the two major notions that give access to the overall topology of the picture, the algorithm can be describe as follows. We consider the picture as a landscape initally covered by the sea, the level of which will be slowly decreased. Three main events can then occur. (i) a new island appears, (ii) an existing island grows, (iii) two nearby islands merge. As far as the topology of the picture is concerned, only the first and last events are important, as they describe a qualitative change in the description of the picture: a new island or a new super-island appears. This, and only this, has to be taken into account. This information is of topological nature. It can be summarized in a tree structure. A new leaf is formed when a new island appears, branching occurs when two islands merge. This is what describes the dendrone in fig.1b. It is essentially coding the way stable subregions of the picture connect to eachother as the scale increases, from the luminance point of view, leaving out all geometrical and morphological details. We nevertheless incorporate some labelling information which will be used for later refinements. The vertical axis represents the intensity levels. The horizontal axis has no specific meaning. In particular it does not describe any ordering.

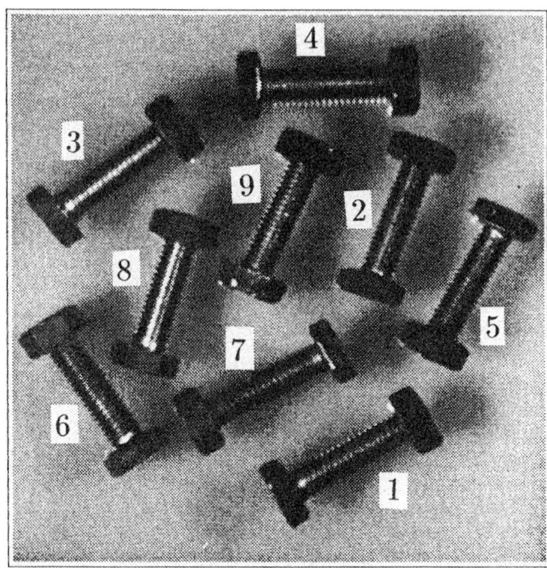

Figure 3: A real world picture with 9 bolts

As a partial coding of the picture content, the dendrone is very robust to noise perturbations, since the overall topology of picture regions is not very sensitive to the amplitude of local noise. Only very small objects, the size of which is close to noise correlation length, are significantly perturbated. This robustness property is demonstrated in fig.2, where the noise amplitude is greater than that of the disks. The dendrone nevertheless clearly shows the existence of "two objects" as two persistent subdendrones.

The dendronic structure is not only compact and robust, but it can be also processed to improve the detection of objects. Various filtering algorithms can be applied to it to increase its significance content. One can for instance analyse the distribution of branches lengths to detect "sticking out" parts. In this case, the dendrone is used as a pure structural representation of the data, only the vertical axis being used to build a metric space. Note that ultrametric distances can also be defined on such a tree. But, as already mentioned, the dendrone can be labelled with some geometric information such as the position, surface area or elongation factor of the islands that each node represents. More refined algorithms can use this information. For instance, from the spatial autocorrelation function of the picture, one can measure the local amplitude of noise and its correlation length. One can then define a noise correlation volume, which will be used to filter out from the dendrone all branches having a volume, defined from the branch height and base surface area, lower than this correlation volume. In this way, the definition of noise can be obtained from the picture or the dendrone itself, and used to process the dendrone. This is shown in the next example with a real world picture. Figure 3 represents a set of nine bolts on a more or less uniform background but with noticeable shadows and noise. In fig.4a is shown

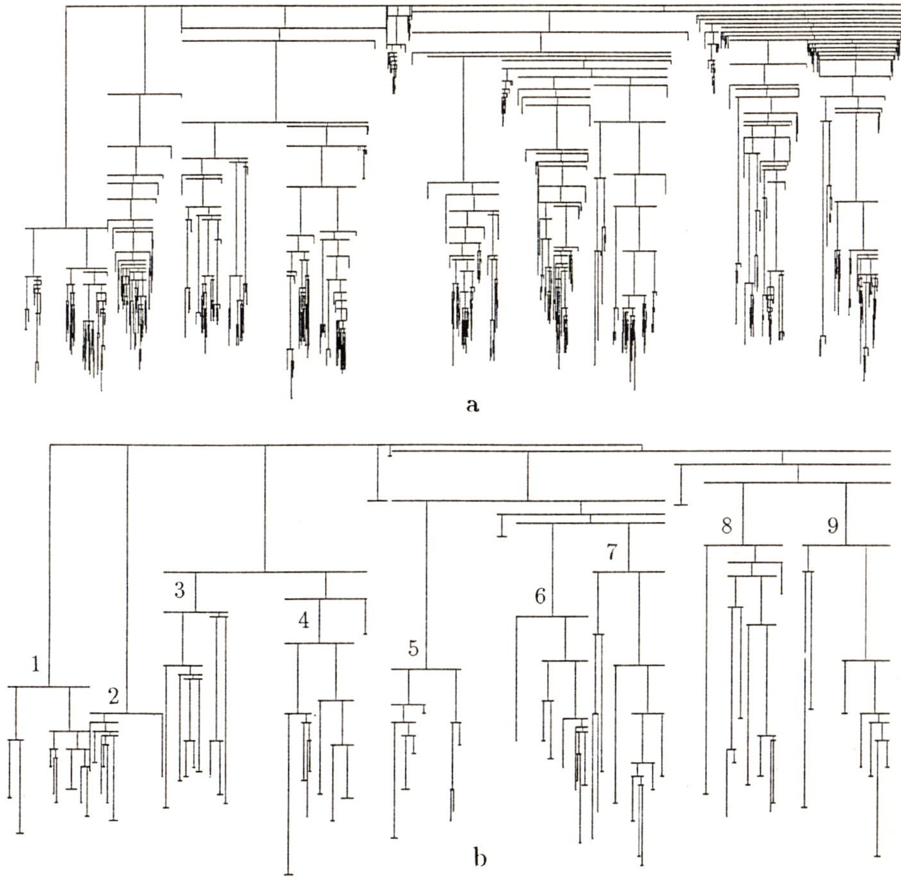

Figure 4: Dendrone of fig.3 (a) Filtered dendrone (b) horizontal width of subdendrones proportional to island surface area

the corresponding dendrone, with a particular representation in which the width of each subdendrone, featured by an horizontal segment, is proportional to the surface area of the base of the corresponding island. In this way, the horizontal scale is used to represent some interesting labelling information.

In fig.4b is shown the result of filtering using the noise correlation volume. In other words, all dendrone nodes leading to too small isotropic branches are eliminated from the dendrone. This does not mean cutting out branches, but rather replacing a forked branch with a single branch. This is applied recursively to the dendrone, with amounts to replacing some subdendrones by a single branch, so defining a coarse grained view of the picture, keeping only the main branching points. Increasing the range of filtering, by considering a larger cutoff volume –not presented here–, one could get a dendrone with only nine branches corresponding to the nine objects which are present in the picture. In this filtering one

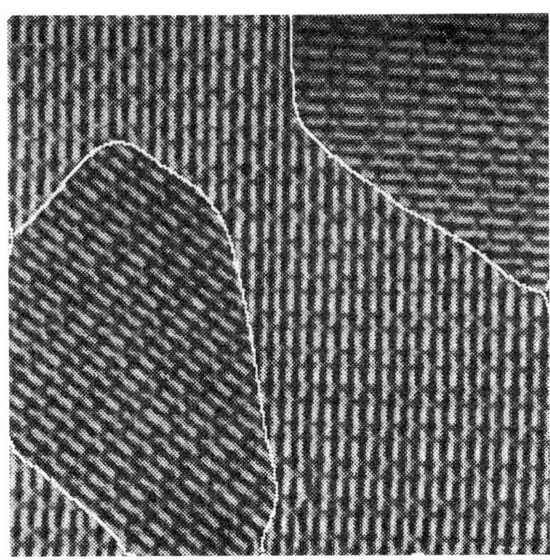

Figure 6: Result of segmentation (boarder detection) of textures by inverse Gabor filters driven by the dendrone of the Fourier transform

the end points of the curve. This is clearly not good enough. We can refine this view by looking for the point, along the arc spanned by this segment, which is the furthest away. We can then define a two-segment approximation by including this new point in the description. The same process is repeated recursively on each arc. In this procedure each segment is splitted in two, leading to a better approximation. This defines a branching process from which a binary tree can be constructed. If each node in this tree is labelled with the distance at which the corresponding segment was entered in the description, as accuracy is increased, we obtain the dendrone we are looking for. In fig.7b is given the dendrone of the Koch curve in which the vertical axis is in logarithmic scale. From top to bottom the accuracy increases, and correspondingly, the number of segments required in the description, at that accuracy, increases. The dendrone is very regular, including only one branch height in logarithmic scale, which simply reflects the existence of only one scale factor. Obviously the dendrone exhibits the construction rule of the Koch curve.

It is interesting to see how the amount of information included in the description increases as a function of accuracy requirements. In fig.7c is plotted the logarithm of the number of segments, i.e. the number of branches in the dendrone, as a function of the logarithm of the distance. Its envelop is a straight line the slope of which is exactly the fractal dimension of the curve ($\log(4)/\log(3)$).

Let us consider now a less simple case. In fig.8a in drawn a Koch curve with two scales factors ($1/2, 1/3, 1/3, 1/2$). The corresponding dendrone is given in

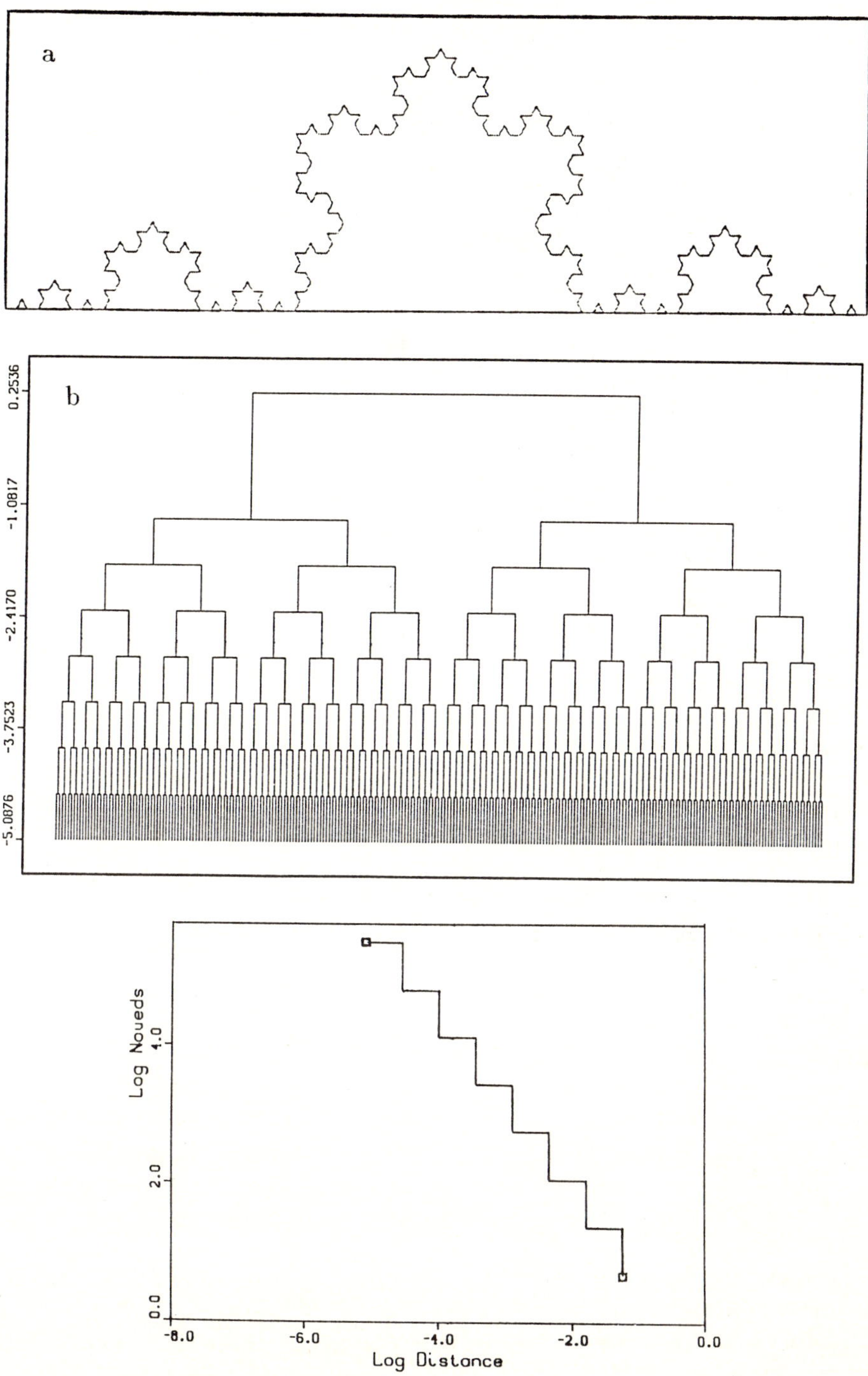

Figure 7: Regular Koch curve (a) its dendrone (b) and the number of segments vs accuracy in logarithmic scale (c)

fig.8b. The structure is clearly visible. A simple inspection can reveal again the rule of construction, and the presence of only two scale factors. A clearer view can be obtained if one considers a 1 to 4 or two-step branching. On can then see that there is only one stable pattern with the two expected factors. This analysis of the dendrone can easily be made automatically, which turns out to be necessary for more complex cases. One can detect the motives that repeat regularly and their structure. The same method can be applied to 4-factor multifractal Koch curves, the dendrone of which cannot be analysed visually, and the exact four factors along with the rule of construction can be exhibited.

In fig.8c is plotted the number of segments as a function of accuracy. Again a straight envelop allows us to define a fractal dimension. But, as for other fractal dimension definitions and measurements producing such a graph, it is clear that it defines only a global measure of fractality, reflecting a uniform behaviour as a function of scale. For such a multi-fractal curve, the dendrone contains more information, in fact all what we have to know about the data.

The fact that the Dendronic Analysis is able to capture so much information results here again from its ability to be driven by the data, achieving an adaptability which produces semantic information, another example of the *respect* of the signal. Contrary to other methods such as box counting fractal measurements or wavelets [WAV89], one does not fix any particular diadic grid in position and scale, or try to look for a scale which is not present in the object. Structural elements or scales that are present are directly extracted.

4. Trajectories of Dynamical Systems

Extending this approach to higher order objects, one could analyse lines in any number of dimensions, such as the trajectories of a dynamical system. One of the questions which arise in this context is to establish a correlation between the complexity of the dynamical model and that of the behaviour, represented by the trajectory. We are thus lead to analyse the morphology of a trajectory and to quantify the amount of information which is required to describe it.

To this aim we can perform a Dendronic Analysis of the trajectory in the complete phase space. A typical situation is depicted in fig.9, in which the number of segments in again plotted as a function of the accuracy. We detect here two plateaus, which reveal the existence of two stable descriptions. Along a particular plateau no information needs be introduced as we ask for more accuracy in the description, hence the term "stability" of description, or persistence, that we have already used in the context of pictures. We see here that the Dendronic Analysis is able to detect the internal levels of description of the observation.

In many instances, the amount of morphological information required within a plateau is surprisingly small, leading to a tremendous data compression. Only essential features are preserved, and the analysis, through this plot, is able to tell us what "essential" means. From this reduced view of the trajectory it is then possible to analyse the contributions of the various processes in the dynamical model and simplify it by detecting, for instance, negligeable parts or coupled contributions.

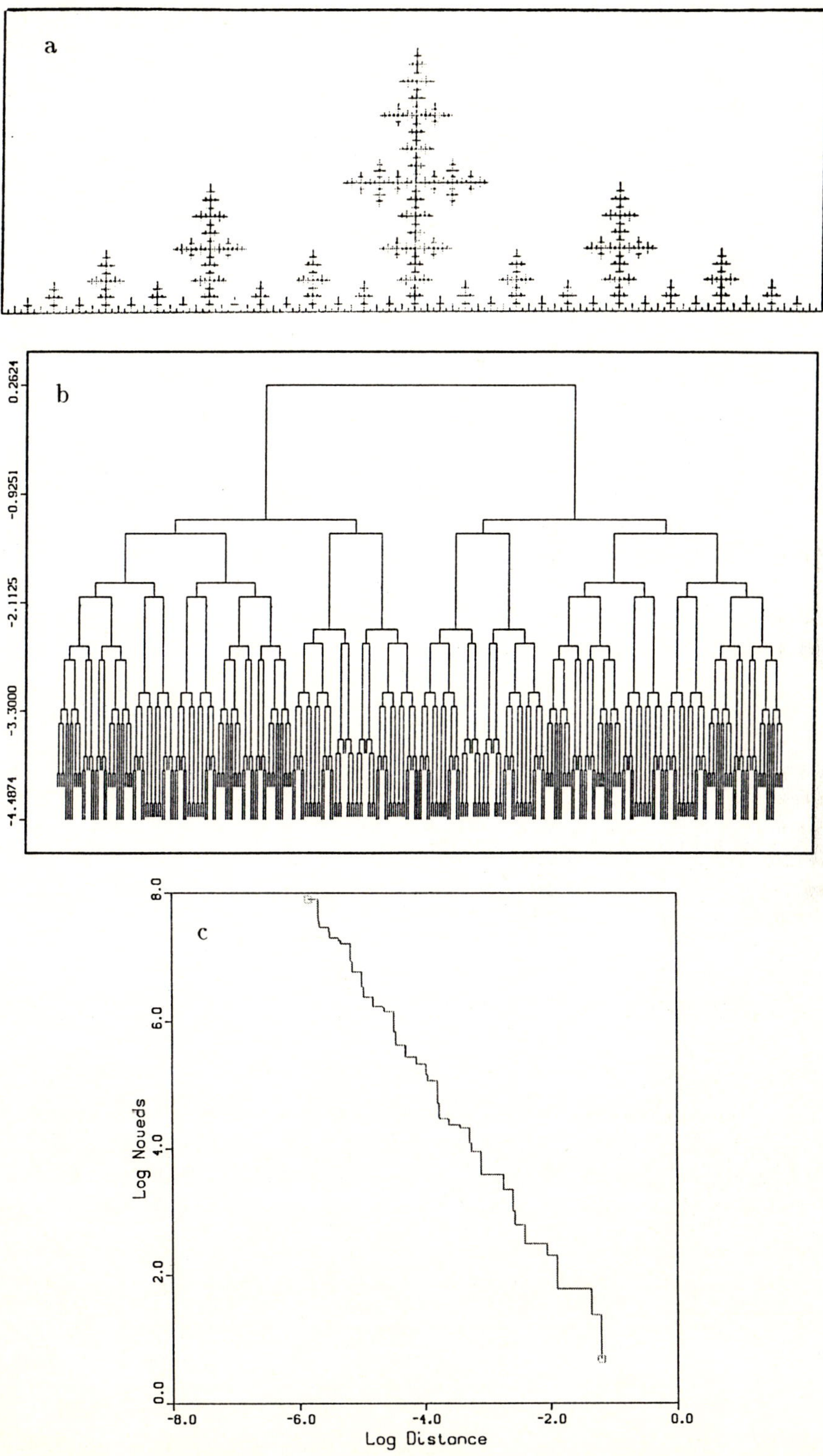

Figure 8: Koch curve with 2 scale factors (a) its dendrone (b) and the number of segments vs accuracy in logarithmic scale (c)

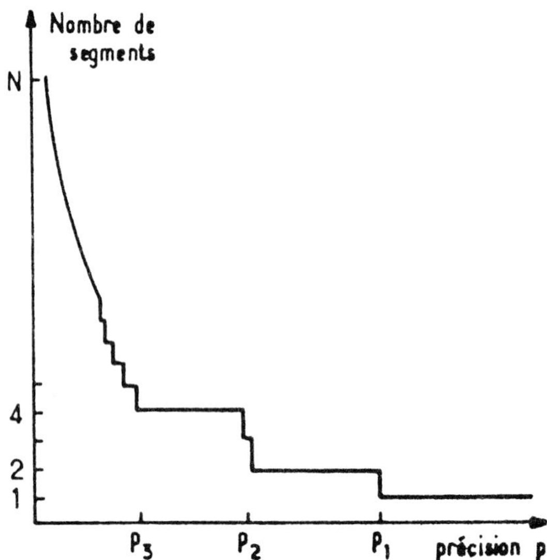

Figure 9: Number of segments vs accuracy for a trajectory having two stable levels of description, one for each plateau

This was applied to the chemical model shcwn in fig.10a, in its complete form, including 13 variables (dimension of phase space) in 33 processes [HAN87]. In a typical situation, it produces a limit cycle oscillation. Analysing the trajectory along one period allows us to derive a reduced version of the model (fig.10b), given underneath, still in chemical notation, which includes only 4 variables in 7 processes. The quantitative agreement, note described here, is remarkable. Further reduction could be achieved but at the expense of loosing the ability to keep a chemical type formulation, which is still valid from the dynamical point of view, but to some extent unpleasant, for a chemist at least.

This particular example shows an instance of dendrone-driven symbolic treatement of a model. The complete, realistic, but obscure initial formulation is transformed into a compact, more tractable and understandable one.

5. Conclusion

This set of examples demontrates the ability of the Dendronic Analysis, the goal of which is to produce in a data-driven way a tree structure, the Dendrone, which is coding essentially the semantic content of the signal. A number of valuable properties are exhibited, such as robustness, context-free detection, invariance under euclidian geometrical transformations.

One of major features of the approach, as a general methodology for complexity analysis, resides in its ability to detect the internal levels of description of an observation. The way information enters in the description as a function of accuracy requirements provides self-definition of these levels. We have seen two

```
CSTR KO Br-, BrO3-, ArOH, H2O, H+

BrO3-  + Br- + 2H+    = HBrO2 + HOBr       ; 2.1, 1.D+4   (R4,R5)
HBrO2  + Br- + H+     = 2HOBr              ; 2.D9         (R6)
HOBr   + Br- + H+     = Br2 + H2O          ; 8.D9, 110.   (R7,R8)
BrO3-  + HBrO2 + H+   = 2BrO2. + H2O       ; 1.D4, 2.D7 (R9,R10)
2HBrO2                = BrO3- + HOBr + H+  ; 4.D7         (R11)
BrO2.  + ArOH         = ArO. + HBrO2       ; K1           (R12)
BrO2.  + BrArOH       = BrArO. + HBrO2     ; K2           (R13)
Br2    + ArOH         = BrArOH + Br- + H+  ; K3           (R14)
HOBr   + ArOH         = BrArOH + H2O       ; K4           (R15)
BrArO. + ArO.         = DH + Br-           ; K5           (R16)
BrArOH + ArO.         = BrArO. + ArOH      ; K6,K7        (R17,R18)
```

a

```
X=Br-, Y=HBrO2, Z=BrO2.  et W=ArO.  :

X        = Y           ; (P1)
Y + X    =             ; (P2)
Y        = 2 Z         ; (P3,P4)
2 Y      =             ; (P5)
Z        = W + Y       ; (P6)
W        = 1/2 X       ; (P7)
```

b

Figure 10: A dynamical system defined by a realistic chemical reaction scheme (a) After dendrone-driven analysis of its trajectory a reduced model is produced (b)

typical cases. In the last example, the existence of plateaus, reveals the presence of a stable description, its counterpart in pictures being the sticking out character of a particular sub-dendrone. In the case of fractals or multifractals, a linear dependence in logarithm scale, reveals uniform scaling properties. Other types of classifiable or generic behaviours could be defined.

As a semantic oriented approach, the Dendronic Analysis can be applied to symbolic treatements. The last application can be considered as an example. Nevertheless, one of the foundation of this approach resides in the use of a metric space, which most physical measurements provide, in with some distance has to be defined and used. We are thus able to move from a numeric level to a symbolic one. Each level of description posseses its objects and their inter-relationships, in other word its proper language.

Could we, with the same ideas, process language itself, or symbolic spaces, strings of symbols? The questions are presently under investigation in the context of theroretical linguistics.

References

[HAN87] P. Hanusse: Analyse morphologique de trajectoires en dynamique chimique, J.Chim.Phys. 84,1315-1327, 1987

[HAN90] P. Hanusse, Ph. Guillataud: Object detection and identification by hierarchical thresholding, Lectures notes in Computer Science, 427, Computer Vision ECCV 90, O. Faugeras Ed., Springer 1990 ECCV90

[HAN91] P. Hanusse, Ph. Guillataud: Semantique des images par Analyse Dendronique, Proceeding of the RFIA conference, Lyon, AFCET Ed. Paris, November 1991

[SAM87] H. Samet, R.E. Webber: Hierarchical data structures. A review", Proceeding of the International Electronic Image Week, Nice, April 1986, CESTA Paris 1987

[TAN80] S. Tanimoto, A.Klinger: Structured Computer Vision, Academic Press 1980

[WAV89] Wavelets, L.M. Combes, A. Grossmann, P. Tchamitchian Eds, Springer 1989

Texture Analysis Using Fractal Dimensions

U. Müssigmann

Fraunhofer Institute for Manufacturing Engineering and Automation

1 Introduction

The ideas and methods of the fractal geometry to characterize physical processes and structures which show scaling behaviour have attracted interest in many sciences [MAN83].

One interesting application of fractal geometry is the field of image analysis, especially texture analysis. The apparent rough structure of many natural surfaces which is reflected in a corresponding roughness of the grey value surface of an image suggests to apply methods from fractal geometry for an analysis. Naturally, such an approximate identification with a fractal makes sense only within certain scale boundaries.

The basic idea for the development of a recognition and classification algorithm based on this interpretation consists in using the concept of fractal dimension (such as, for example, Hausdorff dimension, Minkowski dimension and metric dimension) as a property characterizing the scaling behaviour of a fractal.

However, it is well known that different sets with the same fractal dimension may differ substantially in their structure (figure 1).

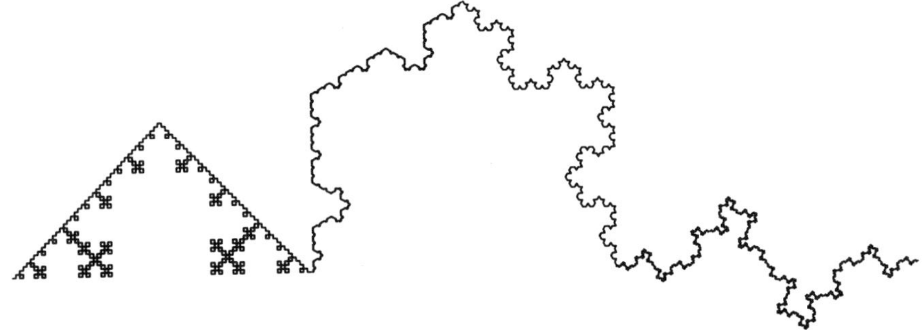

Fig. 1: Inhomogeneous fractal set

One possibility to handle this circumstance is to describe a set A not only by a fractal dimension but by an infinite number of different generalized dimensions $D(q)$ $(q > 0)$ with $D(q) \geq D(q')$ for $q' > q$ ([BAD85], [GRA83], [FED88]).

If the equality holds for all q, the fractal is called homogeneous [HEN83]. For a non-uniform fractal the dimension function $D(q)$ takes different values. In this case the fractal is called a multi-fractal [VOS88].

In this paper we propose a completely different characterization of a multi-fractal set which makes use of the well-known concepts of the Hausdorff dimension.

For this purpose we first define the local Hausdorff dimension. With this definition we are able to prove several characteristics of multi-fractal sets.

The concept of a local fractal dimension can also be applied to texture classification and segmentation. The objective of texture segmentation is the decomposition of an image into homogeneous regions (figure 2).

Fig. 2: Textile with defects

Identifying the textural homogenity with homogenity with respect to the fractal dimension of the set of grey values, this decomposition can be achieved by calculating the local fractal dimension for each image point.

A major problem in analysing fractal sets is the development of algorithms for the estimation of a fractal dimension. In this paper we present a new method for the calculation of a fractal dimension of textures with the help of the scale space filtering.

2 Local fractal dimension

When wandering around in different parts of a set S, the degree of fractality may change appreciably. Our approach for a quantitative characterization of such an inhomogeneous fractal set S will be based on partitions of S into subsets S_i which are homogeneous with respect to the Hausdorff dimension.

Definition Let (S, d) be a metric space. A subset T of S is called homogeneous with respect to the Hausdorff dimension, if for every $x \in T$ and $\epsilon > 0$ the following equivalence holds

$$dim\big(B(x,\epsilon) \cap T\big) = dimT$$

(with $B(x, \epsilon)$ being the closed ball with radius ϵ and center x).

Let S be a set with metric d we then consider partitions

$$S = \bigcup_{i=1}^{N} S_i$$

with the properties

1) $S_i \cap S_j = \emptyset \quad (i \neq j)$
2) $dimS_i \neq dimS_j \quad (i \neq j)$
3) $x \in \partial S_i \cap \partial S_j \ , \quad dimS_i > dimS_j \quad \Rightarrow \quad x \in S_i$ (with ∂S_i being the boundary of S_i)
4) The subsets S_i are homogeneous with respect to the Hausdorff dimension.

Having given this definition, we are confronted with the following questions. Does there always exist such a partition for an arbitrary set S and, if that holds, is it unique? Furthermore, which values take the Hausdorff dimensions of the homogeneous subsets S_i? In order to proceed we define a local Hausdorff dimension.

Definition Let (S, d) be a metric space and x an element of S. The local Hausdorff dimension of x in S is defined as follows:

$$dim(x, S) := \lim_{\epsilon \to 0} dimB(x, \epsilon) \ .$$

With this definition we are able to define an equivalence relation on S; two points x, y in S are called equivalent if they have the same local Hausdorff dimension,

$$x \equiv y \ :\Longleftrightarrow \ dim(x, S) = dim(y, S) \ .$$

(the proofs of the following lemmas and theorems are given in [MUE90]).

Lemma 1 Let x be an element of an equivalence class T defined by the local Hausdorff dimension with $x \in \partial T$. If T' is another equivalence class with $T \neq T'$ and $x \in \partial T'$ we have

$$dim(x, S) > dim(y, S) \quad \text{for every } y \in T' \ .$$

Further considerations will be based on the assumption that the set S has the following properties:

- Let S be a metric space with countable base, i.e. every open subset of S is the (countable) union of sets of the base.
- There is only a finite number of equivalence classes defined by the local Hausdorff dimension.

For such a restricted set S the first theorem can be formulated.

Theorem 1 Let T be an arbitrary equivalence class of S defined by the local Hausdorff dimension, then

$$dim(x, S) = dimT \quad \text{for all } x \in T \ .$$

This means that knowing the local Hausdorff dimension of only one point of T we immediately get the Hausdorff dimension of the whole equivalence class T. This obviously is a very nice property of the local Hausdorff dimension.

In addition, there holds the following theorem.

Theorem 2 The equivalence classes defined by the local Hausdorff dimension are homogeneous with respect to the Hausdorff dimension.

We can prove the following lemma.

Lemma 2 Let $\mathcal{C} = S_1, \ldots, S_M$ be an arbitrary finite cover of S, $S = \bigcup_{i=1}^{M} S_i$, satisfying conditions 1) – 4). If we have $x \in S_k \in \mathcal{C}$, then

$$dim(x, S) = dim S_k \ .$$

From the Theorems 1 and 2 and Lemmas 1 and 2, it follows that the partition of the set S defined by the equivalence classes is the unique partition of S into homogeneous subsets.

3 Fractal analysis

The mathematical definition of the Hausdorff dimension requires to consider various limits which are hard to perform in practice. Therefore, many other definitions of fractal dimensions were introduced, e.g. information dimension, correlation dimension and capacity dimension, together with algorithms to calculate them (see e.g. [YUI86]).

To characterize the scaling behaviour of a fractal set and thus to estimate a fractal dimension it is necessary to analyse the fractal on several scales. This is done by analysing a sequence of smoothed versions of the given signal. Examples for such smoothings are sausages or parallel sets for the calculation of the Minkowski dimension (figure 3) and polygons for the "Coastline-of-Britain" analysis (figure 4).

Fig. 3: Parallel sets of the Koch curve

Fig. 4: Approximation of the Koch curve using polygons

Experimental investigations show that the following power law is valid for the length $L(\epsilon)$ of the polygons with edge length ϵ:

$$L(\epsilon) \propto \epsilon^{1-\text{fractal dimension}} .$$

The fractal dimension can be estimated by calculating the logarithm of the length of the polygons in dependence on the logarithm of the edge length which is shown for the Koch curve in figure 5.

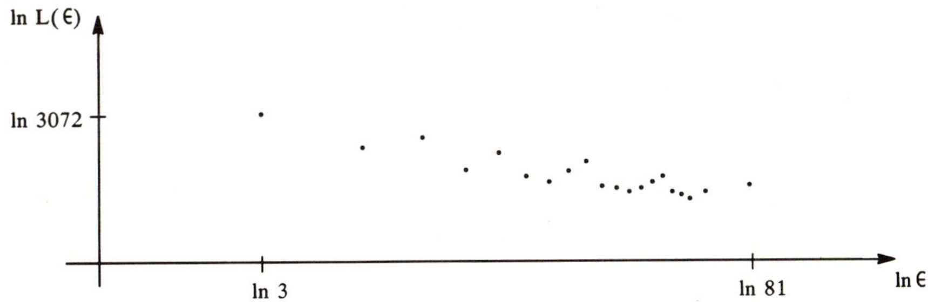

Fig. 5: Richardson plot

The diagram shows that the logarithm of the length of the polygons does not always fall onto a straight line.

Another way to smooth a signal in order to estimate its length makes use of the so-called "scale space filtering" which was developed originally in pattern recognition to describe an image qualitatively by its extrema. The idea underlying the scale space filtering can also be used to describe the fractal behaviour of a signal. The smoothing is done by a convolution of the signal $I(x)$ with a Gaussian kernel $g(x, \sigma)$ with standard deviation σ,

$$C(x, \sigma) = I(x) * g(x, \sigma) = \int_{-\infty}^{\infty} I(u) \frac{1}{\sigma\sqrt{2\pi}} e^{\frac{-(x-u)^2}{2\sigma^2}} du .$$

This convolution defines a continuous sequence of smoothed functions in dependence on the standard deviation of the Gaussian (figure 6).

Fig. 6: Sequence of Gaussian smoothings of the Koch curve

As in the case of the Richardson algorithm a scaling exponent can be calculated by computing the length $L(\sigma)$ of the smoothed signal $C(x, \sigma)$. The scaling exponent can be estimated as the slope of the straight line we get via regression of $\ln L(\sigma)$ over $\ln \sigma$ (figure 7).

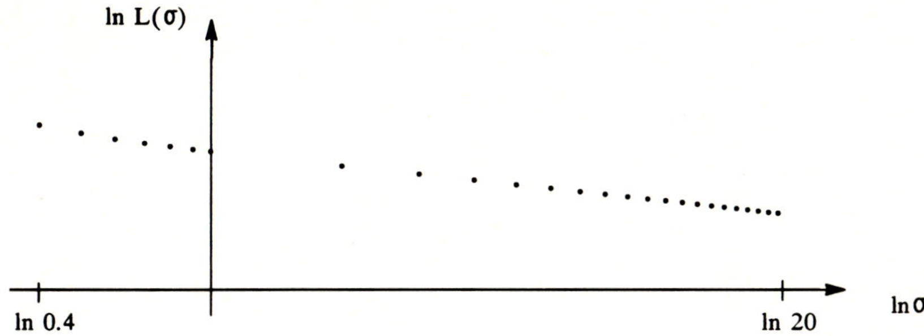

Fig. 7: Length of Gaussian smoothings in dependence on standard deviation

The proposed method to analyse the scaling behaviour using a Gaussian filter function can be extended to the analysis of two-dimensional signals. Given a signal (or an image) $I(x, y)$, we calculate the surface area F in dependence on the standard deviation σ,

$$F(\sigma) = \int \sqrt{1 + C_x(x, y; \sigma)^2 + C_y(x, y; \sigma)^2} dxdy ,$$

with C_x and C_y being the partial derivatives of the smoothed signal C of I ([MUE89]).

Again, the scaling exponent can be estimated via regression of $\ln F(\sigma)$ over $\ln \sigma$.

4 Texture analysis

Based on the method of scale space filtering, we have applied the concept of a local fractal dimension to texture segmentation. As examples we have chosen discrete grey value images G,

$$G = [0, \ldots, N-1] \otimes [0, \ldots, M-1] \rightarrow [0, \ldots, I-1] ,$$

which are composed of two regions with different textures. The proposed method of segmentation results from the identification of textural homogenity with homogenity with respect to the fractal dimension of the set of grey values. Thus we estimate the local scaling exponent of each image point using scale space filtering by taking into account some neighborhood of the point. The scaling exponent obtained in this way is then assigned to the respective point. The result of this first step is a "dimension image". The last step for texture segmentation is to combine all points with approximately the same fractal dimension to one region. In our examples the images were composed only of two textures. In these cases we thus end up with a binary image. All points that belong to the first texture take the value 0, the other points 1.

In the following we present the resulting segmentation of different images. Figure 8 shows the first texture. It is composed of different self-affine fractal structures generated by computer simulations. The figure shows the original image (left), the dimension image (middle) and the result of the segmentation.

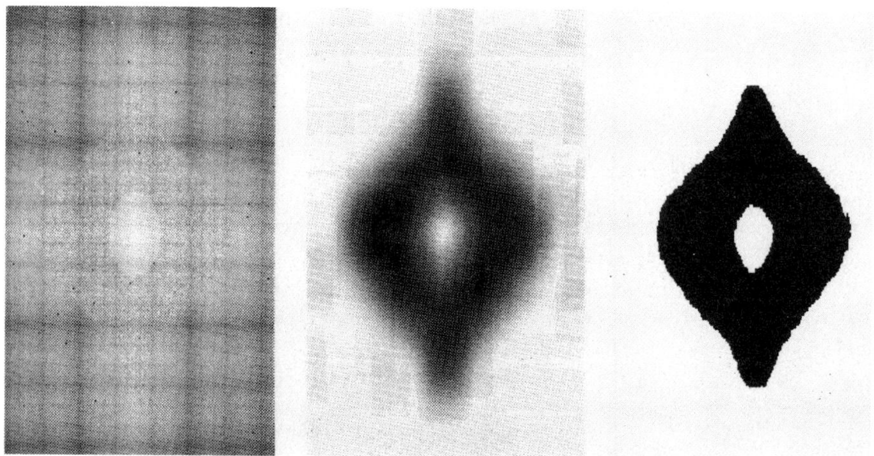

Fig. 8: Segmentation of a computer generated test image (self-affine structure)

Figure 9 shows the corresponding result for a real texture taken from Brodatz.

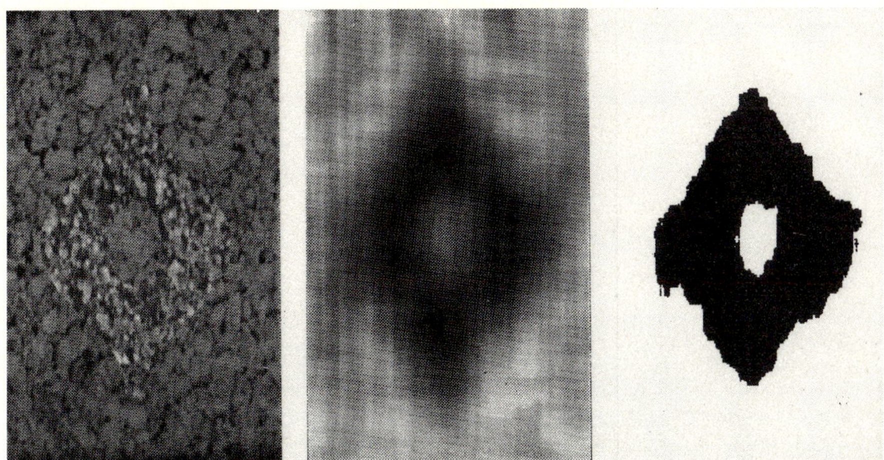

Fig. 9: Segmentation of different Brodatz textures (D33, D29)

Texture analysis methods are very often used for the automation of quality inspection tasks in industry. One typical application is the detection of defects in natural surfaces as shown in figure 2. The resulting segmentation using the scaling exponent as a feature is shown in figure 10.

Fig. 10: Detection of surface defects

Other examples of industrial applications are given below.

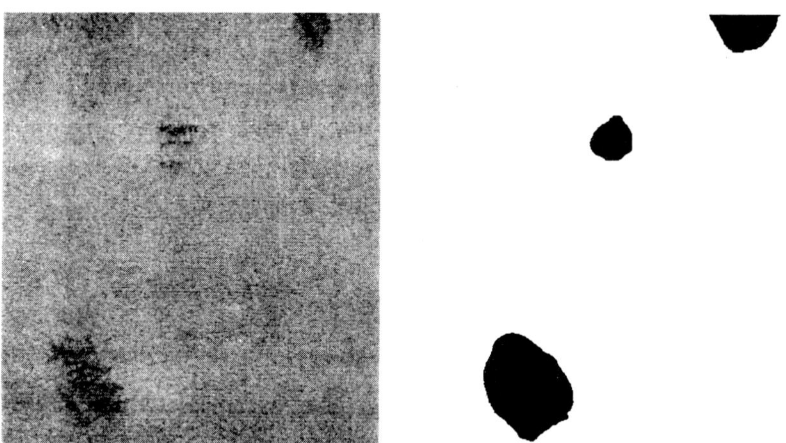

Fig. 11: Detection of surface defects (leather)

Fig. 12: Natural surface with defects (stone)

Fig. 13: Detection of surface defects

5 Summary

Starting from the concept of a local Hausdorff dimension we have proved several basic theorems which show that this concept is a useful tool for the characterization of multi-fractal sets. We applied these ideas to the problem of texture segmentation whereby we identified textural homogenity with homogenity with respect to the fractal dimension. By use of a new method to estimate the local fractal dimension of an image point we obtained very promising results.

6 References

[BAD85] R. Badii, A. Politi, "Statistical description of chaotic attractors: The dimension function", Journal of Statistical Physics 40 (1985), 725 – 750

[FED83] J. Feder, "Fractals", Plenum Press New York (1988)

[GRA83] P. Grassberger, "Generalized dimensions of strange attractors", Physical letters 97A (1983), 227 – 230

[HEN83] H.G.E. Hentschel, I. Procaccia, "The infinite number of generalized dimensions of fractals and strange attractors", Physica 8D (1983), 435 – 444

[MAN83] B.B. Mandelbrot, "The fractal geometry of nature", W.H. Freeman, San Francisco (1983)

[MUE89] U. Müssigmann, "Texture analysis, fractals and scale space filtering", Proc. 6th Scandinavian Conference on Image Analysis, Oulu, Finland (1989)

[MUE90] U. Müssigmann, "Homogeneous fractals and their application in texture analysis", Proc. Fractal '90, 1st IFIP Conference on Fractals, 1990

[VOS88] R.F. Voss, "Fractals in nature: From characterization to simulation", in The Science of Fractal Images, eds. H.O. Peitgen und D. Saupe, Springer-Verlag (1988)

[WIT83] A.P. Witkin, "Scale-Space Filtering", Proc. Intern. Joint Conference Artif. Intell., Karlsruhe (1983)

[YUI86] A.L. Yuille, T.A. Poggio, "Scaling Theorems for Zero Crossings", IEEE Patt. Anal. Mach. Intel., Vol. 8, No.1, 1986

Limited Selfsimilarity

H. R. Bittner

Institut für Biochemie und Endokrinologie, Gießen

Due to influence of biggest and smallest details of natural objects, fractal measurements can yield selfsimilarity only within a certain range. A log-logistic function is used to describe both fractal and topological ranges. This leads to both differential fractal dimensions (scaling exponents), and possible scaling residues.

1 Problem

Measuring natural objects with biggest and smallest details at varied resolutions leads to fractal properties within certain limits. Thus selfsimilarity can be found only asymptotically in a confined range of scale. If the description is restricted to the selfsimilar range only, which may be far from the limits, the influence of all values outside this range, which can be measured with similar precision, is neglected. Depending on the actual task, this effect could be negligible, or considerable. In order to include the topological and the intermittent range for the description of architecture with small and medium range of selfsimilarity as well, the use of a log-logistic fitting function is proposed, taking into account also a fractal or topological *scaling residue.*

2 Calculation of the Fractal Capacity Dimension

When determining fractal dimensions D_F by box counting technique [MAN82], technical requirements or experimental restrictions posed by the measuting method may cause that the relationship of box number N vs. characteristic length ε cannot be measured as original abscissa (x) or ordinate (y) values.

Thus when taking into account both the dimensions D_x, D_y of the units of the abscissa and ordinate resp., the general form for the relations of D_F and the slope b in the log-log-plot is found by:

$$D_F = D_y - b \cdot D_x \qquad b = \frac{D_y - D_F}{D_x} \tag{1}$$

This generalized form is nowise trivial and especially useful for the case of measuring fractals by fractals.

This is the case, for example, in measuring the porosity of gels by size exclusion as a function of the molecular weight MW of the penetrating test molecules (Fig. 1). Here the dimension of the ordinate, the pore volume, is $D_y = 3$, and the abscissa MW is itself a fractal (dimension D_x), since for globular proteins, the molecular weight is not proportional to the third power of the mean hydrodynamic molecular radius MR, but to $MR^{2.62}$. For oblong polyethylenglycols (PEGs) as testmolecules, the relation is even lower ($MW \propto MR^{1.94}$) [SER89].

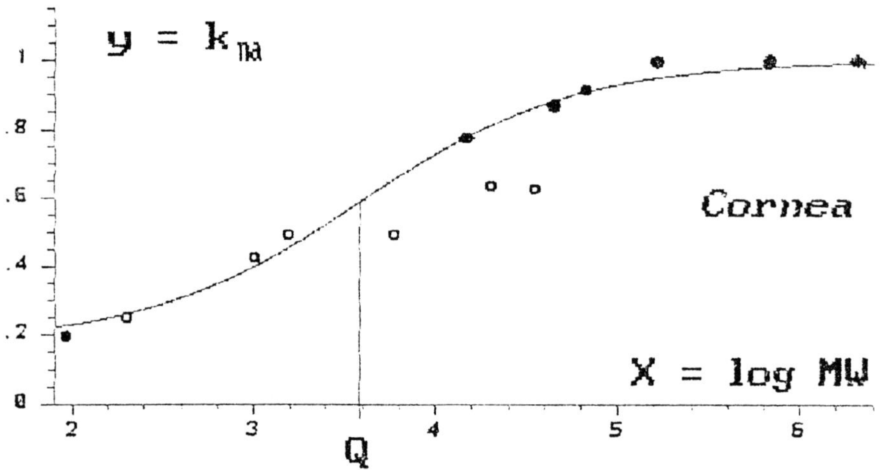

Fig. 1: Porosity measurement at the cornea: Relative volume k_{na} inaccessible to test molecules of molecular weight MW [SER91], [BAC91]. o : PEGs, • : proteins and glycerol. Matrix fraction $y_{min} = 0.18$

3 Selfsimilarity Within Limits

When fractal analysis is applied to real structures, the sizes of the smallest detail (lower limit) and of the biggest detail or of the entire object (upper limit) cause a changing scaling behaviour, hence leading to a deviation from an ideal, unlimited linear log-log-plot.

We encounter the problem of confined selfsimilarity in chromatographical tasks, for instance, in determining the porosity of a network of cross-linked polymer chains by gel filtration. Here the accessible volume for dissolved molecules of different size as test elements is measured, ranging say from 100 dalton to 10^6 dalton. The dependence ends at the magnitudes of the smallest and the biggest pores. Outside this range, changes in molecular size have no effect on the accessible volume fraction k_{av} or its complement k_{na}. In the log-log-plot of the inaccessible fraction k_{na} vs. MW (Fig. 1), the sigmoid form of the resulting cumulative pore size distribution yields a quasi-linear region only in that range, where the porosiy shows nearly selfsimilar properties [SER88], [SER90].

If the description is restricted to data of the selfsimilar range only, which may be far from the limits (in y), by definition of a voluntarily chosen cutoff in X at a "critical resolution", the influence of all values outside this range, which can be measured with similar precision, is neglected. Since the transition towards the limits is smooth, the question arises:

Which range on the abscissa is necessary to allow fractal evalution?

We show that by introducing the log-logistic function the entire range of scale can be used for fractal description [SER89].

4 The log-logistic Function (LLF)

The *log-logistic function* describes the sigmoid course of the ordinate y as a function of the logarithmic abscissa X = log ε between the limits y_{min} and y_{max}. It can be developed from the selfsimilar power function $y = a\varepsilon^b$, $\varepsilon > 0$, by substitution of its derivative by an expression with upper and lower limits:

$$\frac{dy}{dX} = y \cdot b \Longrightarrow \frac{(y_{max} - y)\,(y - y_{min})}{(y_{max} - y_{min})} \cdot b$$

$$y = y_{max} - \frac{y_{max} - y_{min}}{1 + e^{b(X-Q)}} = y_{min} + \frac{y_{max} - y_{min}}{1 + e^{-b(X-Q)}} \tag{2}$$

$$\frac{dy}{dX} = (y_{max} - y_{min})\,\frac{b \cdot e^{b(X-Q)}}{(1 + e^{b(X-Q)})^2} = \frac{(y_{max} - y)\,(y - y_{min})}{(y_{max} - y_{min})} \cdot b$$

with Q = parameter of position, where $y = \frac{1}{2}\,(y_{min} + y_{max})$, X = ln ε (note the changed sign in the second exp-function).

The derivative shows the influence of the distance to the limits, namely (y_{max} – y) and (y – y_{min}), within the range (y_{max} – y_{min}) on the increment dy/dX. It becomes zero in approaching one of the limits.

The LLF is discussed in more detail elsewhere [BIT91a]. It describes properties which are distributed symmetrically along a logarithmic abscissa (Fig. 2a). If symmetry is not necessarily required by theoretical considerations, it has to be proved by experiment. The LLF occurs e.g. at cooperative enzyme konetics or as generalized saturation function or isotherm, with b as an exponent, descibing the intensity of cooperative effects between two interacting components.

The equation in the above form includes the general case of a non-zero lower boundary y_{min} as *scaling residue*, and its coordinates clearly indicate the magnitude and the distribution width of fractal properties [SER89], [BIT89].

With kown limits y_{max} and y_{min} from measured data the parameters b and Q can be obtained either with a nonlinear regression analysis or by linearization after *logit transformation* of the ordinate (Fig. 2c):

$$\text{logit } y = \ln \frac{y - y_{min}}{y_{max} - y} = b\,(X - Q) \tag{3}$$

If the relation does obey a log-logistic function, this transform leads to a straight line of logit y vs. X, a plot, which resembles the common linear graph of unlimited selfsimilarity. Q results as the intersection point with the abscissa and b as the slope of the fitting line.

The *nonlinear regression* to the data from the LLF distinguishes in the different weighting of aberrations near the limits. Whereas the logit transform considers deviations near the limits more grave than in the medium y-range, the nonlinear regression uses equal weight for variations in y.

The decision between the two methods has to take into account the peculiarities of the problem and of the measuring technique, so no general preference exists [BIT91a].

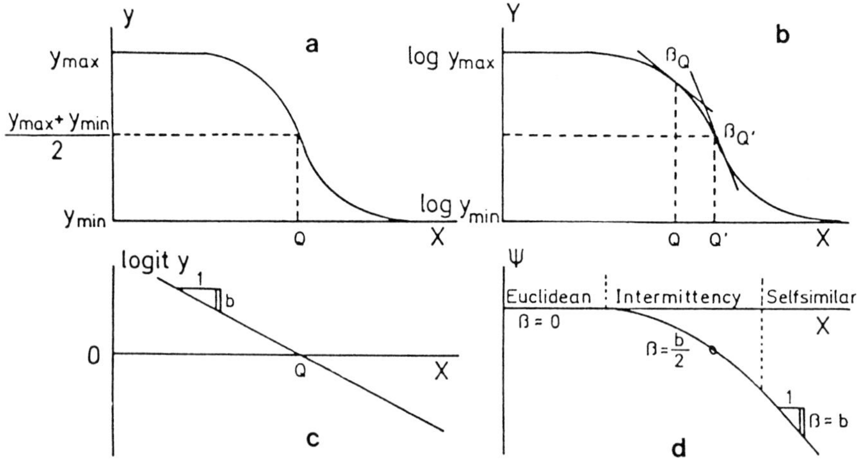

Fig. 2: Measurement of a property y, distributed within a limited range: (β denotes the local slope) **a)** plot with linear ordinate y **b)** data after logit transformation for determination of b and Q **c)** plot with logarithmic ordinate Y = log y **d)** log-log-plot with $\Psi = \log(y - y_{min})$, clearly indicating nearly selfsimilar, Euclidean, and intermittent range

5 Fractal Interpretation

The fractal dimension results from the derivative, the (local) slope β of the log-log-plot according to eq 1. For fractal characterization either the logarithm of the ordinate quantity, $Y = \log y$, or only of that part, exceeding the scale invariant residue y_{min}, namely $\Psi = \log (y - y_{min})$ has to be used (Fig. 2b, d). Here again the decision depends on the peculiarities of the measurement.

For a sigmoid Y(X) relation, usually only the middle part with approximately linear dependence is chosen for fractal evaluation. Since this choice is influenced by individual tolerance, and since the nearly linear range may be narrow, the calculation of the scaling exponent may lead to differing results.

Fitting an LLF to the Y(X) dependence yields a scaling exponent $\beta = dY/dX$, varying from an (usually) topological, integer value to the selfsimilar scaling exponent and back again to topological values.

$$Y = \log y \qquad = \log \left[y_{max} - \frac{y_{max} - y_{min}}{1 + e^{b(X-Q)}} \right]$$

$$\Psi = \log (y - y_{min}) = \log \frac{y_{max} - y_{min}}{1 + e^{-b(X-Q)}} \tag{4}$$

For a vanishing residue y_{min}, Y and Ψ are equivalent. The maximum steepness of Y(X) occurs at the inflexion point $Q^* = Q + (\ln R) / 2b$, with $R = y_{min} / y_{max}$.

$$\beta (Q^*) = \left. \frac{dY}{dX} \right|_{X=Q^*} = \frac{1 - \sqrt{R}}{1 + \sqrt{R}} \cdot b \tag{5}$$

The transition from limited to unlimited selfsimilarity ($y_{min} = 0$ and $y_{max} \rightarrow \infty$) leads to the fractal case with b as dimensional excess, whereas Q, growing with log y_{max}, loses its meaning, since a prominent range of scale no longer exists.

The LLF describes the continuous fading of selfsimilar properties towards the limits. Here the scale dependence, given by the scaling β, need not be constant, and hence the scaling exponent itself is a function of the resolution and leads to a *differential fractal dimension $D_F(X)$*. Depending on the actual problem, the scaling $\beta = D_T - D_F(X)$ is given either by dY / dX or by $d\Psi / dX$ [BIT89].

At porosity measurements of a matrix network of cross-linked chains (Fig. 1), Y has to be used for the determination of the network capacity dimension. Ψ in contrast shows the scaling behaviour of the inaccessible volume, caused by the network, but disregarding the volume of the network itself [SER90]. The following result is obtained for the scaling β:

$$\beta (X) = D_T - D_F(X) = \frac{d\Psi}{dX} = b / (1 + e^{b(X-Q)}) \tag{6}$$

For approaching the upper limit ($e^{b(X-Q)} >> 1$), the object behaves nearly topologically ($D_F(X) \approx D_T$). For $e^{b(X-Q)} << 1$, the scaling of Y again approaches topological behaviour (i.e. the dimension of y_{min}), whereas the scaling of Ψ approaches the exponent b, i.e. the almost selfsimilar case.

It should be noted, that the LLF fitting and interpretation described here, is different from the concept of *multifractals*, where a sigmoid $D_q(q)$ relation occurs [TÉL88]. In our case, the selfsimilarity limitation of a geometrical measure is described, whereas in multifractals a predominant (fractal) subset is chosen by certain 'cost functions'.

The *distribution* of fractal properties results from the derivative dy/dX of the integral quantity y.

$$\frac{dy}{dX} = (y_{max} - y_{min}) \frac{b \cdot e^{-b(X-Q)}}{(1 + e^{-b(X-Q)})^2} \tag{7}$$

Due to normalization, the parameter b, which in the selfsimilar power function represents the dimensional excess, here becomes the measure of *dispersion* as well. In the case of porosity, it determines the width of the pore volume distribution [BIT91a].

The limits y_{max} and y_{min} need not be topological parameters, like length or accessibel volume. The above considerations are also valid for fractal limits. Then y_{min} is a fractal scaling residue, and the LLF shows the transition between different ranges of selfsimilarity. Figure 3a-c screens several cases of superposition of an LLF (f_1, $y_{min} = 0$) with an unlimited power function. The combination of two unlimited power functions resembles the Σ graph in Fig. 3c, since the course of the f_1 function at negative X values has only little effect on the sum [BIT91a].

By coupling several LLFs and power functions, a variety of Σ graphs can be yielded. For practical relevance in terms of fractal description, of course, each contributing functional has to be argumented. For instance, superposing some LLFs with equal b and harmonic Qs and y_{max}s (with $\Delta Q = 3$ and subsequent y_{max}-ratio of 2) may result in a power-law-like dependence with overlayed harmonic oscillations (Fig. 3d).

Such kind of relation may be found at anatomical and physiological investigations, e.g. at the diameter vs. branching generation on the lung [WES86], or at understanding the organism in terms of geometry or blood dynamics as sum of all organs, whose contributions are superposed [SER90].

The LLF enables us to utilize data from both tails of the distribution for fractal analysis. The LLF is suited only for the description of properties which are distributed symmetrically along a logarithmic scale abscissa. Before application of the LLF, this symmetry has to be proved by experimental data, or at least has to be made plausible [BIT91a].

Structures with small and medium range selfsimilarity, where the effect of limits is considerable, can be found regularly in nature, especially in biosciences. Examples are the already mentioned porosity of gels, other cross-linked polymer networks, and biological tissues [SER91], the allometric phenomena in comparative physiology (reduction law of metabolism) [SER90] (seen at mammals below 100g body mass), and the branching of blood vessels [BIT91b]. In automated image analysis the lower limit might be important at geological or material science testing [BIT89].

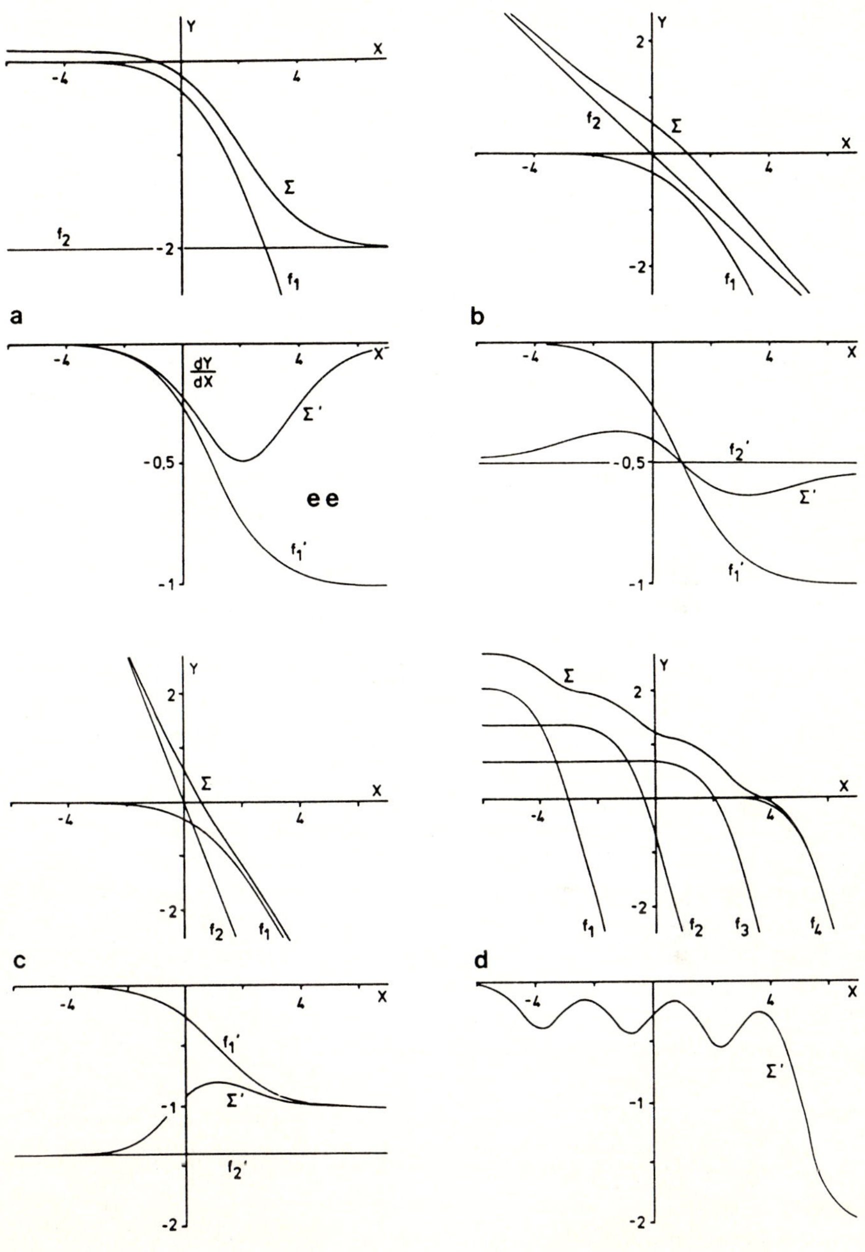

Fig. 3: Superposition of two functions f_1, f_2 with different scaling behaviour.
Values (top) and derivatives (bottom) are shown for f_1, f_2 and for the sum $\Sigma = f_1 + f_2$
f_1: $y_1 = (1 + e^{-x})^{-1}$ (i.e. $b = -1$, $Q = 0$, $y_{min} = 0$)

a) f_2: $y_2 = \text{const}$

b+c) f_2: $y_2 = e^{bX}$ ($b = -0.5$ and -1.4)

d) four LLFs with $b = -2$, $Q = 5 / 2 / -1 / -4$, $y_{max} = 1 / 2 / 4 / 8$

Moreover the LLF may serve to interprete cooperative interactions known in antigen-antibody or enzyme kinetics, O_2-binding of haemoglobin, in fractal properties of the underlying system [SER89].

The differential $D_F(X)$ expand the characterization of fractal structures from the strictly selfsimilar range to the entire range of non-Euclidean behaviour with changing fractal properties.

Acknowledgement

This work was supported by grant Se315/13-3 of the Deutsche Forschungsgemeinschaft. The author is on leave from the above institute. He wishes to thank Prof. Manfred Sernetz for the fruitful cooperation and intensive discussions in the past years.

References

[BAC91] P. Bach: Mikrointerferometrische Bestimmung der Porenvolumenverteilung von Cornea und Kontaktlinse und ihre fraktale Interpretation. Thesis, Gießen 1991

[BIT89] H.R. Bittner, P. Wlczek, M. Sernetz: Characterization of Fractal Biological Objects by Image Analysis. Acta Stereologica, 1989, 8, 31-40

[BIT91a] H.R.Bittner, M. Sernetz: Selfsimilarity within Limits – Description with the Log-Logistic Function. In: H.-O. Peitgen, J.M. Henriques and L.F. Penedo (Eds.): FRACTAL'90 – Proc. of the 1st IFIP Conference on Fractals. Lisbon, June 6-8, 1990, Elsevier, Amsterdam 1991 (in press)

[BIT91b] H.R. Bittner: Modelling of Fractal Vessel Systems. In: FRACTAL'90 (see above), 1991

[MAN82] B.B. Mandelbrot: Fractal Geometry of Nature, Freeman, New York, 1982

[SER89] M. Sernetz, H.R. Bittner, H. Willems, C. Baumhoer: Chromatography. In: D. Avnir (Ed.): The Fractal Approach to Heterogeneous Chemistry. J. Wiley, 1989, 361-379

[SER90] M. Sernetz, H. Willems, H.R. Bittner; Fractal Organization of Metabolism. In: W. Wieser, E. Gnaiger (Eds.): Energy Transformations in Cells and Organisms. Thieme, Stuttgart 1990, 82-90

[SER91] M. Sernetz, H.R. Bittner, P. Bach, B. Glittenberg: Fractal Characterization of the Porosity of Organic Tissue by Interferometry. In.: F. Rodriguez-Reinoso, K.S.W. Sing, J. Rouquerol (Eds.): Characterization of Porous Solids II (COPS II, Alicante, May, 1990), Elsevier, Amsterdam 1991, 141-150

[TÉL88] T. Tél: Fractals, Multifractals, and Thermodynamics, Z. Naturforschung 43a, 1988, 1154-1174

[WES86] B.J. West, V. Bhargava, A. Goldberger: Beyond the Principle of Similitude: Renormalization in the Bronchial Tree. J. Applied Physiology 60(3), 1986, 1089-1097

Fractal 3D Analysis of Blood Vessels and Bones

P. Wlczek*, A. Odgaard♦, M. Sernetz*
*Justus-Liebig University Giessen
♦Orthopaedic Hospital Aarhus

1 Introduction

In nature we often observe highly complex structures, which cannot be described by euclidean geometry [MAN82]. These objects show the phenomenon of resolution dependence of measures e.g. surface. Using the width of the resolution range information about the scaling properties can be derived in form of a scaling exponent. Fractal geometry offers the possibility to interpret this exponent as a fractal dimension. The dimension represents the intensity of the folding. We use the concept of fractal geometry to describe the branching system of the arterial kidney vessels and the structure of trabecular bone.

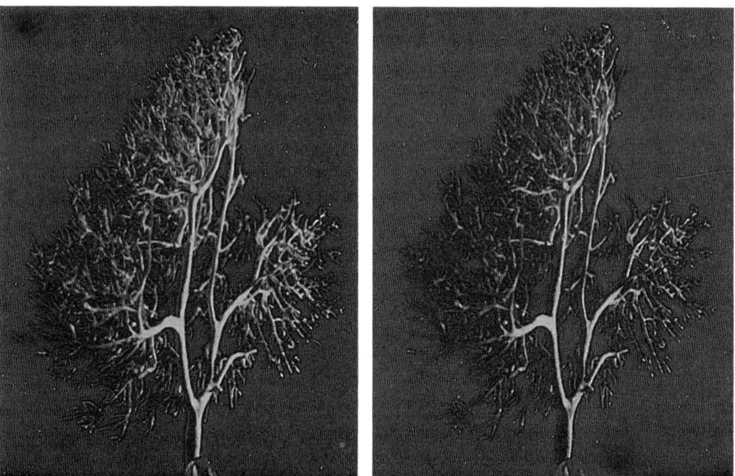

Fig. 1: Stereoscopic image of a coloured plastic cast of the vessel system produced by the corrosion casting technique. By this technique the vessels of the kidney are filled with coloured polymer resin. After the plastic is hardened the tissue is removed by alkali

2 Methods

2.1 Three-dimensional data-sets

In order to obtain three-dimensional data-sets for the application of fractal geometry we use two methods, namely mechanical preparation of serial sections with a turning lathe [WLC89] or a microtome [ODG90] and nuclear magnetic resonance imaging. The three-dimensional data-sets which were produced by NMR-Computertomography could be used directly for the fractal analysis.

For the mechanical preparation the casts in turn were embedded in white plastic. With a lathe such plastic cylinders were cut into sections of 0.1 mm thickness. The surfaces of each cut were photographed by a CCD-camera which was fixed on the axis of the lathe. After each cut a binary picture was stored on a PC. From this stack of pictures the three-dimensional data-sets of the vessel system were reconstructed. A similar method was used for the three-dimensional data-set of the bones. In contrast to the described mechanical preparation of serial sections with a lathe here a hard tissue microtome was applied to cut the plastic block.

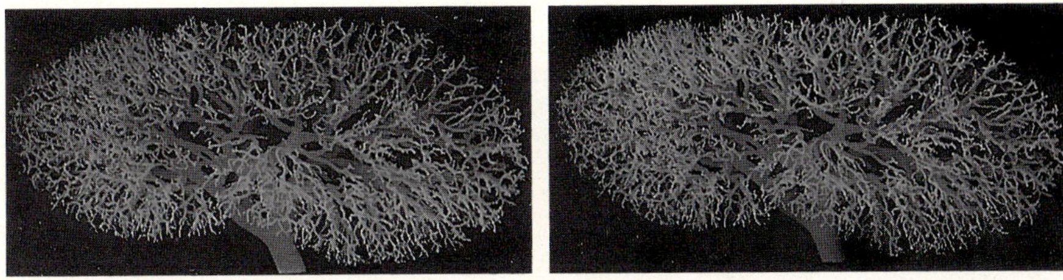

Fig. 2: Stereoscopic image of a vessel system reconstructed from mechanical serial sections of an arterial kidney cast. The maximum length of the object is 105 mm

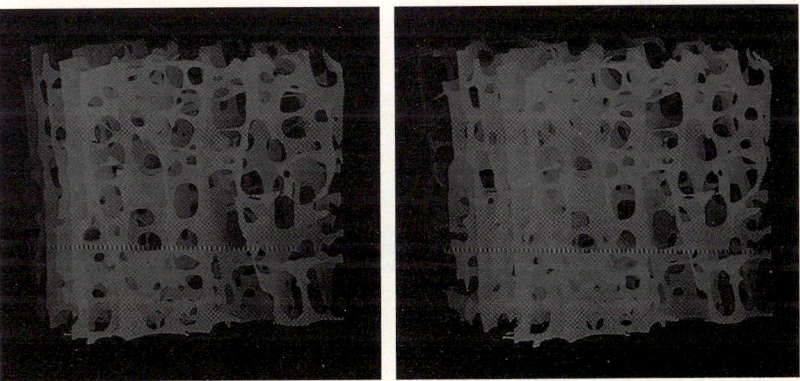

Fig. 3: Stereoscopic image of trabecular bone reconstructed from serial sections. The maximum length of the object is 7 mm

With the described methods it is possible to produce three-dimensional data-sets with a maximum resolution of 20 μm. For fractal analysis it is necessary to describe the scaling properties over a wide range of scale. Therefore it is not only important to achieve data-sets with high resolution but it is also necessary to gain a great field of view of the structured object. The mechanical preparation of serial sections offers the possibility to achieve data-sets with a field of view of 512 pixels in x and y direction. The number of sections is unlimited in z direction.

2.2 Analysis of scaling behaviour

Various analysing methods exist to determine the fractal dimension [VIC89]. The common approach of all methods is to inspect the object with different resolution. We use the mass-radius analysis to measure the scaling of the mass as a function of the resolution.

At first the mass-radius analysis will be explained by means of a plane fractal object (skeleton of a coral, Fig. 4). Then the method will be extended to the three-dimensional case.

Fig. 4: Binary picture of the skeleton of a coral superimposed with circles of increasing radii for mass-radius analysis

To determine the fractal dimension with the mass-radius analysis an arbitrary point of the data-set is chosen as a center of circles with different radii. For each radius the number of points on the circle area is determined (Fig. 4).

The different diameters simulate different resolutions. For a selfsimiliar object, as in our case, the number of points as function of radius scales according to a power law. From the slope of the regression line in a double logarithmic plot the exponent respectively the fractal dimension can be calculated.

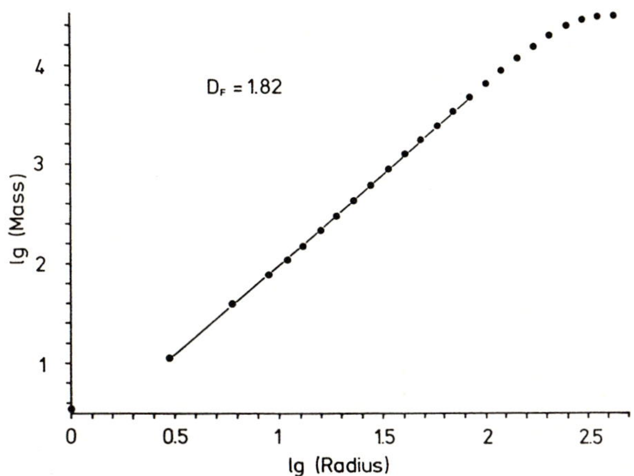

Fig. 5: Double logarithmic plot Mass M versus Radius R of the structure in figure 4. From the slope of the regression line a fractal dimension of 1.82 was calculated

For the three-dimensional data-sets of the kidneys and the trabecular bones the described mass-radius analysis was extended and spheres instead of circles were used. For natural and computer generated objects we have to take into account, that these objects have limited size and in contrast to mathematical constructions are not infinitely iterated. Therefore natural objects have a scaling behaviour only in a definite scaling range limited by an upper and lower cut off.

Since the mass-radius relation is sensitive to the position of the center of the spheres different positions have to be chosen to calculate the average number of points in a sphere of defined radius. In the best case all points of the data-set have to be used as a center.

3 Results

3.1 Fractal analysis of 3-D data-sets

Up to now six data-sets of kidneys and two of trabecular bones were evaluated by the described mass-radius analysis. Additional to the whole mass, also the surface of the data-sets was analysed. For this purpose data-sets were derived which contain only the voxels on the surface.

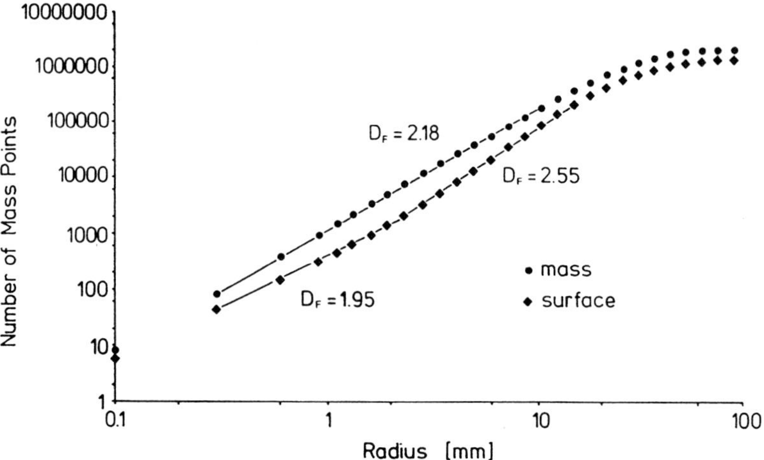

Fig. 6: Double logarithmic plot of the number of mass (•) and surface (◆) points versus the radius for kidney 5. For the calculation 110 center points were chosen at random

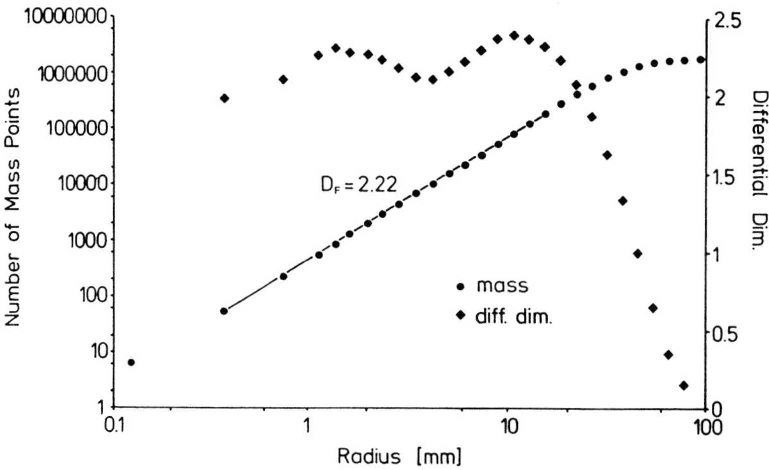

Fig. 7: The differential dimension versus the resolution of data-set kidney 6. The differential dimension has been calculated for each point by linear regression between the two adjacent points

In figure 6 the slope of the regression line for the mass yields a fractal dimension of 2.18. The evaluation of the surface of the data-set yields two distinct scaling ranges. The first at high resolution gives a dimension of 1.95 and the second at low resolution gives a dimension of 2.55. As typical for any natural fractal this plot shows the influence of the upper and lower cut offs. For radii of 20 mm and above the slope of the curve decreases to zero. This effect is due to the limited extension of the object. For very large radii or low resolution the structure shrinks to a point.
For the interpretation of the fractal analysis we additionally use the differential dimension. Figure 7 shows that in the range of 0.4 mm to 20 mm the differential dimension varies between 2.1 and 2.4. Corresponding to the evaluation of the mass the differential dimension decreases to zero for large radii.

Table 1 shows mass and surface dimensions together with the evaluated range of resolution of the kidney and bone data-sets.

Dimensions of data-sets

Name of data-set	Dim. of mass	Dim. of surface
Kidney 2	2.20 (0.45-22mm)	2.05 (0.45-3.75mm) 2.47 (4.65-22.35mm)
Kidney 3	2.19 (0.45-15mm)	--
Kidney 4	2.35 (0.38-18mm)	2.06 (0.38-2.38mm) 2.47 (2.88-21.87mm)
Kidney 5	2.18 (0.30-10mm)	1.95 (0.30-2.8mm) 2.55 (2.80-14.6mm)
Kidney 6	2.22 (0.38-15mm)	--
Bone 1 A10	2.08 (0.06-0.46mm) 2.58 (0.56-3.5mm)	2.08 (0.06-0.46mm) 2.64 (0.56-3.5mm)
Bone 2 A17	2.19 (0.06-0.46mm) 2.56 (0.56-3.5mm)	--

Table 1: Mass and surface dimensions together with the evaluated range of resolution of the kidney and bone data-sets

In case of the kidney the mass-radius analysis yields different behaviour of scaling for the mass and for the surface. For the mass we find an average dimension of 2.2. The evaluation of the surface yields two distinct scaling regions, the first ranges from 0.4 mm to 3 mm with a dimension of about 2.02, the second ranges from 3.4 mm to 20 mm with a dimension of about 2.5.

In contrast to the kidney the bone structure shows two distinct scaling regions in both cases. The first range reaches from 60 μm to 460 μm with a dimension of 2.08 respectively 2.19, the second reaches from 560 μm to 3.5 mm with a dimension of 2.57.

3.2 Modelling

We have used fractal geometry not only for the description of the morphology of the vessel system, but in turn we applied it also for the construction of three dimensional vessel models by recursive computer algorithms (Fig. 8) [BIT89]. The design parameters of the branching system are vessel length, cross-section and branching angle. The construction on the computer allows to compare both morphological and kinetic properties of model and original (Fig. 9).

Fig. 8: Stereoscopic image of a simulation of a vessel system constructed by recursive computer algorithms

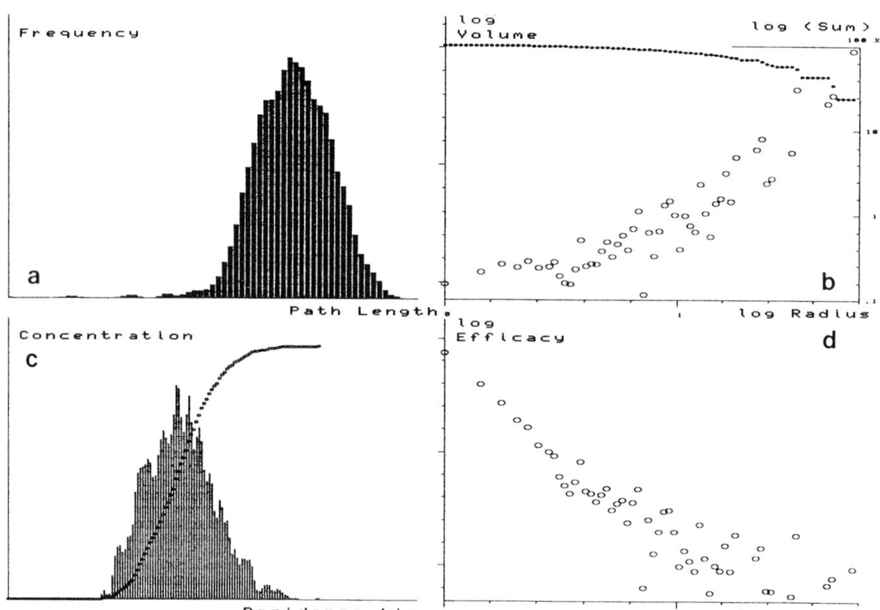

Fig. 9: Morphological and dynamical parameters, computed from the model in Fig. 8: a) Distribution of path lengths from basis to tips b) Distribution (•) and scaling (°) of vessel volume c) Residence time distribution (e.g. of a blood cell) and integral (°) d) Distribution of physiological efficacy (e.g. oxygen exchange)

4 Discussion

The interpretation of the branched vessel system as a fractal transport system can be used to explain e.g. the peculiar allometric relationship of metabolic rate to the size of organisms [SER85]. This phenomenon has been known and described since long, but up to now it has not been interpreted satisfactorily.

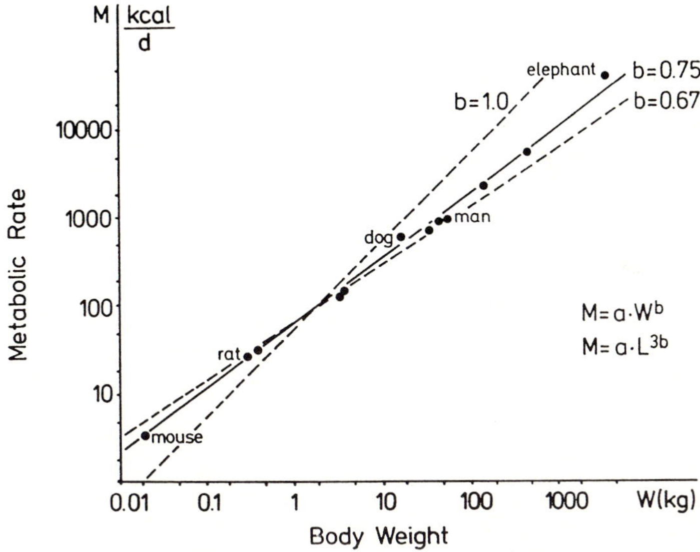

Fig. 10: Relationship between the metabolic rate M and the body weight W in a log log plot

The socalled law of reduction of metabolism (Fig. 10) can be formulated as a power law with noninteger exponent. For the dependence of metabolism on body mass the exponent was measured with 0.75. So the metabolic rate is not directly proportional to the mass respectively to the volume of the organisms with the exponent of 1 and neither to an euclidean surface with an exponent of 0.66. Instead we interpret that exponent as an expression of the fractal structure of the organisms with an exponent of 2.25 for a length scale. Our analysis of the kidneys yield values for the dimension of about 2.20 and supports this assumption.

Therefore the fractal interpretation of branched vessel systems yields a clue to the interpretation of transport limited metabolic processes in organisms and thus to the comprehension of allometric dependence of metabolic rate on body size.

5 References

[BIT89] H.R. Bittner, P. Wlczek, M. Sernetz: Characterization of Fractal Biological Objects by Image Analysis, Acta Stereologica, 8: 31-40, 1989

[MAN82] B.B. Mandelbrot: Fractal Geometry of Nature, Freeman, New York, 1982

[ODG90] A. Odgaard, K. Andersen, F. Melsen, HJ.G. Gundersen: A Direct Method for Fast Three-Dimensional Serial Reconstruction, Journal of Microscopy, 159 3: 335-392, September 1990

[SER85] M. Sernetz, B. Gelleri, J. Hoffmann: The Organism as Bioreactor, Journal of theoretical Biology, 117: 209-230, 1985

[VIC89] T. Viczek: Fractal Growth Phenomena, World Scientific, Singapore, 1989

[WLC89] P. Wlczek, H.R. Bittner, M. Sernetz: 3-Dimensional Image-Analysis and Synthesis of Natural Fractals, Acta Stereologica, 8/2: 315-324, 1989

V. Part: Working Group Results

Random Fractals

Working Group Results

There are two major aspects of the application of random fractals: the first is the modeling and characterization of natural forms and processes; while the second concerns the practical issue of computer simulation of such phenomena. In the first case one is interested in a physically based, accurate model that reflects all relevant aspects of the phenomenon and is in good agreement with theory and experiment. The goal is to improve understanding of the natural phenomenon in order to make predictions possible. Such modeling and analysis is of prime concern to many scientific areas dealing directly with nature (physics, chemistry, biology, geology, astronomy and so on).

The goal of the second case is the direct imitation of nature. A physically accurate model, based on, and in good agreement with, theory and experiment, is rarely necessary. The primary requirement is computational speed and efficiency with "realistic" output. A typical example is the generation of landscape images from fractal terrain models. Although the images look real, the geological time evolution of a landscape is quite different from the fractal approximation. Thus, this second aspect is primarily relevant to computer science in general, and computer graphics in particular.

Consequently, different types of fractal models may be of interest in the two cases. In recent years fractal techniques have received increasing attention in many different scientific areas. The results of this research have yielded a vast amount of literature, which is not always easy to follow. We, therefore, try to identify several main issues for current and future research in both of the above categories.

1 Characterizing Nature

a) Fractional Brownian Motion (fBM) is the model most used for characterizing shapes and processes with a fractal appearance. Indeed, fBM seems to be the most wide-spread method of any type for modeling many natural processes. It has the lure of being both easy to describe mathematically and possessing a substantial literature and tradition of implementation. A consequence of its popularity is that the study of alternative fractal models has been neglected. However,

- What random fractals exist beyond fBM for characterizing natural phenomena?
- For any fractal model, what is the relationship between the model and the corresponding natural phenomenon? In particular, within which limits (upper and lower scale limits) does the model remain valid?

b) The use of fractals typically involves visual perception (i.e., a human observer evaluates the results of the process). Such perception is particularly important for the computer graphics problem of generating and selecting "similar looking" fractals. However, the issue of visual perception and evaluation by a human observer deserves more attention. As an example, the scaling of self-affine fractals alters their visual perception dramatically. Thus, the visual comparison of two fractal requires precise definition of the scale and size used during inspection. Additional questions such as

- "What is the smallest change in the fractal dimension that can be typically detected?"
- "How does a change in the scaling factor or other generating parameter, such as lacunarity, effect the perception of a given fractal?"
- "What is the minimum scaling range necessary for a shape to be perceived as fractal or natural?"

are of interest in this context of perceptual psychology.

c) Time varying phenomena, as well as static shapes, show a fractal character. As a typical example one can mention the fractal curve of a stock market price or other economic variable. Such time-variant processes may also be analyzed and characterized with fractal techniques. However,

- How can fractals be used for making predictions regarding economic or industrial developments?
- What is the reliability of such predictions? How can it be improved?

2 Simulating Nature

a) Here, mathematicals models beyond fBM for synthesizing random fractals are needed.

- How can the existing methods for synthesizing random fractals be extended?

Three possible extensions are:

- generalizations of random walks,
- multi-fractals,
- multiplicative processes.

b) The Fast Fourier Transformation (FFT) technique has been one of the most important and useful tools for the generation of high quality random fractals. However, since the application of FFT first requires a good estimation of the spectral density of the fractal, this method can be difficult to apply for synthesis purposes (i.e., without a fractal sample which may be analyzed). Therefore

- an investigation of transformation methods other than FFT is desirable.

c) A second and serious drawback of FFT lies in its structural globality: the parameters of the fractal, and, thus, of its visual appearance are the same everywhere. This deficiency is shared by most fractal models which generate a fractal by using only its global (average) properties. Although it is very easy to change the global properties of a fractal (e.g., its global fractal dimension),

- most methods do not allow local changes. Only a few local methods have been reported in the past.

d) Most algorithms characterize a fractal by means of only two parameters: the fractal dimension and lacunarity. As an extension of the local properties mentioned above,

- methods using more than these two (global) parameters are needed.

Of particular interest for a) - d) above are wavelet transforms since they provide an increase in the number of parameters and the possibility of locally varying the spectral density. Both the mathematical theory and the practical issues for computer implementation are relevant.

e) Time-varying processes with fractional behaviour other than fBM have received little attention in the past.

- Models of time-varying fractals are of great interest for simulating dynamic phenomena and processes such as turbulent motion and economic developments.

A potential application of these models is to enhance "realism" in computer generated animation and speech.

Modeling and Simulation

Working Group Results

1 Preliminary Remarks

It is not possible to unify all the different statements which have been made during the discussion in the spontaneously arranged round table group *"Modelling and Simulation"*, and of this reason we would like to restrict ourselves to a short report on the various aspects of this theme.

Furthermore, the authors do not seek any unification of their points of views. There was no time and no need for any such unification. Instead, we shall present here a synoptic overview of what we believe is essential with respect to the theme in question. All these aspects have been mentioned at some point during the general group discussion. We are conscious of the fact that the following statements are by no means either sufficiently proven or well elaborated. We think that a special seminar on modelling and simulation should be organized.

2 Experiment and Data

Experiments are performed in order to elucidate an expected relationship. For discussion of general aspects, one should be aware of some linguistic traps: in constructivist terms, *facts* (Germ.: Tatbestände) are made (lat.: *facio, faktum*) by ourselves. In a world without actors, no *facts* exist. Facts are not independent of the experimentalist. In this sense, *data* are not *given* (lat.: *dare, datum*) but *taken* out, *captured* (lat.: *capio, captum*). Thus, "capta" would be a more appropriate term for the result of an experiment [WAT88].

3 Relations and Intuition

Scientists search for significant relationships at low or medium levels of complexity. If relations are too complex the attempt is made to split these into several less complex ones.

After finding a relation, one tries to verify it in terms of a theory that describes the development of the structure under examination. The goal is a theory which includes all *facts* and enables predictions which can be tested by other experiments.

3.1 "Quality" of a theory --> Aim: an aesthetic theory

If the theory is based on brief axioms, and without numereous exceptions, it appears *aesthetic*. For instance, Copernikus' theory of planet motion needed less axioms (heliocentric system, circular motion) than the Ptolomeia notion (geocentric system, a variety of "loop" movements).

The term *aesthetic*, difficult to define for scientific theses, includes the notion of something beautiful, pleasing, harmonic. The experimentalist himself (herself) is a product of nature, developed by the evolutionary process, and hence subject to the general rules of nature. Ways of thinking, perceiving and feeling develop in accordance with nature, so that forms and processes of nature are thus considered to be harmonic, aesthetic. In this sense, the aesthetic criterion is indeed meaningful for application to the assessment of a general principle of nature.

The theory of fractals enables the desription of a new class of complex structures with a reduced set of axioms. In this way, a new group of structures can be based on *aesthetic* theories.

4 What is the Difference Between Modelling and Simulation?

4.1 Description of state - Simulation

Simulation tries to generate an effect (a picture) that most resembles nature.

- e.g. clouds and fog
- e.g. patterns such as those to be found in seashells

4.2 Theory of state / Theory of development - Modelling

Modelling tries to *understand* traditional positions.
There are models which have to be treated like simulations in order to aid understanding - this brings us back to experimentation.
Modelling is an attempt to understand and describe reality by:

- including physical laws
- using knowledge and methods from simulation
- e.g. seashell pattern modelling using cellular automata.
  This concept uses the likely method of the real chemical or biological processes which generate these patterns.

4.3 Verification of models

Einsteins's remark to Heisenberg:"It is impossible to include only observable quantities int a theory. On the other hand, it is the theory that decides, what can be observed."
"... and we have to remember that the things we observe, are not the nature itself, but nature exposed to the manner in which we formulate the question." [HEI59]

- Verification of the theory used for modelling by further experiments.
- Experiments to test predictions of the theory, which generate *capta*, or data, which are not taken from the prior experiments which led to this theory.

5 Theory - Reality - Modelling

We create the reality of the object in our mind. Therefore we require a wealth of experience in order to abstract a model from the situation on which we are reflecting.

A model is thus a construction at a very high level of abstraction, based on a combination of ideas. These ideas are mostly well established by experience. Sometimes we have been able to compress (or reduce) human experience to a set of mathematical axioms. One can then formulate a theory which is totally based on these axioms.

But a scientific model is not only a theory, since one also needs the reality (in addition), to which the model refers. In this strong sense one can only speak about a model if there exists an underlying theory.[MES65]

In most cases, there is no proper theory on which a model can be based. Primas and Müller-Herold have instructively summarized the requirements that a scientific theory must satisfy [PRI84].

Besides this understanding of the term "modelling", there exists a much looser, but traditional and common understanding of what denotes the expression "modelling". This involves ideas (which may or not be parts of a theory) relating to special realities which seem to us to be well understood. Thus, *modelling in this common sense* means the combination of such ideas which give us the feeling of being able to *explain* reality, which is created by ourselves.

A model does not describe real movements, neither of matter nor of mathematical objects. Models only describe the possibilities of mathematical objects. People claim that the movement of real objects can be described in this way, neglecting so called measurement errors.

6 Reality - Simulation

Simulation means that the movement (mostly of mathematical objects) appears to us to correspond with the movement of the real objects under consideration.

Since we are concentrating on the similarity between such movements, there might exist different possibilities for creating various movements of mathematical objects , all of which may simulate the *real* movement of matter. In this sense, simulations reflect reality - the reality of the movement.

For simulations it is not necessary that the mathematical objects reflect objects of the real world (see for example the simulations of the creation of mountain by R. Voss in this issue). But of course it is not forbidden for mathematical objects to reflect parts of the properties of real matter (see for example the simulations of the growth of shells by P.J. Plath in this issue).

In this way simulations can serve as a method for testing experimentally (numerically) the validity of models in the commonly-meant sense.

References

[WAT88] P. Watzlawick, F. Kreuzner: Die Unsicherheit unserer Wirklichkeit, Serie Piper, Piper Verlag, München (1988) pp. 39

[HEI59] W. Heisenberg, Physik und Philosophie, S. Hirzel Verlag, Stuttgart (1959).

[MES65] H. Meschkowski, Mathematisches Begriffswörterbuch, B.I. Hochschultaschenbücher 99/99a, Bibliograpgisches Institut, Mannheim (1965) p. 82.

[PRI84] H. Primas and U. Müller-Herold, Elementare Quantenchemie, B.G. Teubner Verlag, Stuttgart (1984) pp.11

Printing: Mercedesdruck, Berlin
Binding: Buchbinderei Lüderitz & Bauer, Berlin

Beiträge zur Graphischen Datenverarbeitung

J. L. Encarnação (Hrsg.): Aktuelle Themen der Graphischen Datenverarbeitung. IX, 361 Seiten, 84 Abbildungen, 1986

G. Mazzola, D. Krömker, G. R Hofmann: Rasterbild - Bildraster. Anwendung der Graphischen Datenverarbeitung zur geometrischen Analyse eines Meisterwerks der Renaissance: Raffaels „Schule von Athen". XV, 80 Seiten, 60 Abbildungen, 1987

W. Hübner, G. Lux-Mülders, M. Muth: THESEUS. Die Benutzungsoberfläche der UNIBASE-Softwareentwicklungsumgebung. X, 391 Seiten, 28 Abbildungen, 1987

M. H. Ungerer (Hrsg.): CAD-Schnittstellen und Datentransferformate im Elektronik-Bereich. VII, 120 Seiten, 77 Abbildungen, 1987

H. R. Weber (Hrsg.): CAD-Datenaustausch und -Datenverwaltung. Schnittstellen in Architektur, Bauwesen und Maschinenbau. VII, 232 Seiten, 112 Abbildungen, 1988

J. Encarnação, H. Kuhlmann (Hrsg.): Graphik in Industrie und Technik. XVI, 361 Seiten, 195 Abbildungen, 1989

D. Krömker, H. Steusloff, H.-P. Subel (Hrsg.): PRODIA und PRODAT. Dialog- und Datenbankschnittstellen für Systementwurfswerkzeuge. XII, 426 Seiten, 45 Abbildungen, 1989

J. L. Encarnação, P. C. Lockemann, U. Rembold (Hrsg.): AUDIUS Außendienstunterstützungssystem. Anforderungen, Konzepte und Lösungsvorschläge. XII, 440 Seiten, 165 Abbildungen, 1990

J. L. Encarnação, J. Hoschek, J. Rix (Hrsg.): Geometrische Verfahren der Graphischen Datenverarbeitung. VIII, 362 Seiten, 195 Abbildungen, 1990

W. Hübner: Entwurf Graphischer Benutzerschnittstellen. Ein objektorientiertes Interaktionsmodell zur Spezifikation graphischer Dialoge. IX, 324 Seiten, 129 Abbildungen, 1990

B. Alheit, M. Göbel, M. Mehl, R. Ziegler: CGI und CGM. Graphische Standards für die Praxis. X, 192 Seiten, 44 Abbildungen, 1991

M. Frühauf, M. Göbel (Hrsg.): Visualisierung von Volumendaten. X, 178 Seiten, 107 Abbildungen, 1991

D. Krömker: Visualisierungssysteme. X, 221 Seiten, 54 Abbildungen, 1992

G. R. Hofmann: Naturalismus in der Computergrahik. VIII, 136 Seiten, 78 Abbildungen, 1992

J. L. Encarnação, H.-O. Peitgen, G. Sakas, G. Englert (Eds.): Fractal Geometry and Computer Graphics. XI, 254 Seiten, 172 Abbildungen, 1992